China and Great Power Responsibility for Climate Change

As American leadership over climate change declines, China has begun to identify itself as a great power by formulating ambitious climate policies.

Based on the premise that great powers have unique responsibilities, this book explores how China's rise to great power status transforms notions of great power responsibility in general and international climate politics in particular. The author looks empirically at the Chinese party-state's conceptions of state responsibility, discusses the influence of those notions on China's role in international climate politics, and considers both how China will act out its climate responsibility in the future and the broader implications of these actions. Alongside the argument that the international norm of climate responsibility is an emerging attribute of great power responsibility, Kopra develops a normative framework of great power responsibility to shed new light on the transformations China's rise will yield and the kind of great power China will prove to be.

The book will be of interest to students and scholars of international relations, China studies, foreign policy studies, international organizations, international ethics and environmental politics.

Sanna Kopra is a post-doctoral researcher at the Aleksanteri Institute, University of Helsinki and the Arctic Centre, University of Lapland, Finland.

Rethinking Asia and International Relations

Series Editor – Emilian Kavalski, Australian Catholic University (Sydney)

This series seeks to provide thoughtful consideration both of the growing prominence of Asian actors on the global stage and the changes in the study and practice of world affairs that they provoke. It intends to offer a comprehensive parallel assessment of the full spectrum of Asian states, organisations, and regions and their impact on the dynamics of global politics.

The series seeks to encourage conversation on:

- what rules, norms, and strategic cultures are likely to dominate international life in the 'Asian Century';
- how will global problems be reframed and addressed by a 'rising Asia';
- which institutions, actors, and states are likely to provide leadership during such 'shifts to the East';
- whether there is something distinctly 'Asian' about the emerging patterns of global politics.

Such comprehensive engagement not only aims to offer a critical assessment of the actual and prospective roles of Asian actors, but also seeks to rethink the concepts, practices, and frameworks of analysis of world politics.

This series invites proposals for interdisciplinary research monographs undertaking comparative studies of Asian actors and their impact on the current patterns and likely future trajectories of international relations. Furthermore, it offers a platform for pioneering explorations of the ongoing transformations in global politics as a result of Asia's increasing centrality to the patterns and practices of world affairs.

For more information about this series, please visit: https://www.routledge.com/Rethinking-Asia-and-International-Relations/book-series/ASHSER1384

Advaita as a Global International Relations
Deepshikha Shahi

China and Great Power Responsibility for Climate Change
Sanna Kopra

China and Great Power Responsibility for Climate Change

Sanna Kopra

Routledge
Taylor & Francis Group
LONDON AND NEW YORK

First published 2019
by Routledge
2 Park Square, Milton Park, Abingdon, Oxon OX14 4RN

and by Routledge
711 Third Avenue, New York, NY 10017

Routledge is an imprint of the Taylor & Francis Group, an informa business

British Library Cataloguing in Publication Data
A catalogue record for this book is available from the British Library

Library of Congress Cataloging in Publication Data
A catalog record has been requested for this book

ISBN: 9781138557604 (hbk)
ISBN: 9781315151113 (ebk)

Typeset in Times New Roman
by Taylor & Francis Books

For Aino and Senni

Contents

Acknowledgements

This book extends a doctoral dissertation that I defended in November 2016 at the University of Tampere, Finland. The assistance of my supervisors Eero Palmujoki and Lassi Heininen was essential to the completion of my dissertation, and I am eternally grateful for their direction, encouragement and support over the years. I am exceedingly grateful to Cornelia Navari as well, who not only accepted the invitation to serve as my opponent at the public defence of my dissertation but has also provided me with valuable guidance ever since. The book has greatly benefitted from the insightful comments of Cornelia Navari and Tonny Brems Knudsen on my chapter in *International Organization in the Anarchical Society: The Institutional Structure of World Order*, edited by them. My thanks also go to Liisa Kauppila, who has taught me a great deal about China and indulged me with many interesting conversations that helped me progress my work. Moreover, I am deeply grateful for the useful comments offered by two anonymous reviewers of my book proposal.

I would additionally like to acknowledge the financial support of the Joel Toivola Foundation that made it possible to rework my dissertation into this book. I also thank Veli-Pekka Tynkkynen for recruiting me to join the Assessing Intermediary Expertise in Cross-Border Arctic Energy Development project funded by the Academy of Finland (project no. 285959), which in due course enabled me to finish the book.

Last, my greatest thanks go to my husband, Juha, who has always stood by me and never questioned my choices, even if they went against his hopes. Juha: no words are enough to express what your love, support and patience have meant and continue to mean to me.

I dedicate this book to my two daughters, Aino and Senni, who mean the world to me. For your sake – and for the sake of your entire generation – I truly wish that the world's great powers live up to their great power climate responsibility, inspire the whole world to fulfil their climate responsibility and to do so *now*.

Numminen, 8 March 2018
Sanna Kopra

1 Introduction

There has been much talk about responsibility in world politics in recent years. In particular, the allocation of responsibility has been central to international climate negotiations, in which the principle of common but differentiated responsibilities has been agreed upon as a guiding principle. As these negotiations have made clear, however, responsibility is a remarkably vague concept, and its meaning in world politics in particular remains altogether uncertain. In negotiations with stakes as high as Earth's climate, a few questions about responsibility thus need to be asked. For example, what *is* responsibility? When it comes to states, for what are they responsible and to whom? On the international stage, who judges responsibility, its assignment and its fulfilment? What do states need to do, or refrain from doing, in order to be viewed as responsible members of international society?

Responsibility has become an especially popular word in speculations about whether the so-called 'rise' of China will pose a risk or an opportunity for the world. Likewise, an extensive body of academic literature has discussed whether China is, or will become, a responsible player in world politics (e.g. Chan 2006; Clark 2014; Deng 2008; Gill 2007; Gill et al. 2007; Patrick 2010; Shambaugh 2013; Xia 2001; Zhang & Austin 2001). Political debate over China's responsibility has been particularly heated in international climate negotiations, where China has been accused of 'being irresponsible' and 'blocking progress' for years on end (e.g. Lynas 2009; Porter 2009; Vidal 2009). From an adjacent angle, academic research on China's climate policy has focused on the country's contributions to international climate negotiations, its climate policy decision-making process, its national interests in climate negotiations and its responsibility for causing climate change (e.g. Chen, G. 2009; Chen 2012; Ella 2016; Gong 2011; Harris 2011; Harris & Yu 2009; Marks 2010; Moore 2011). In both contexts, China's policies have largely been evaluated with a rubric of Western interests and expectations, and too little attention has been paid to China's own notions of responsibility in international climate politics, particularly on what ethical basis the Chinese government considers itself to be responsible, for what, to whom and, above all, why (cf. Chen, Z. 2009; Foot 2001; Jin 2011; Jones 2014; Scott 2010; Yeophantong 2013; Zhang & Austin 2001).

In this book, I investigate China's evolving notions of great power responsibility, both in general and in the particular context of international climate politics.[1] To some extent, China's rise to the status of a great power can be perceived as a typical change in the international order and thus merely another factor that will shape the diplomatic practices of negotiating procedures and rules about specific international issues. However, I presuppose that China's rise to great power status and its increasing engagement in international practices will not only shape the contemporary international order but also generate a transformation of international norms. China is no doubt relevant to discussions of all norms of international responsibility because its rise could facilitate more profound changes in international society. In the context of climate responsibility in particular, China's role is especially central, both theoretically and practically – not least because China is now the world's largest carbon emitter, so presents a tremendous challenge to mitigating climate change and human security around the world. Consequently, China's engagement in international climate politics is imperative, for without its participation, no global effort to combat climate change will succeed. At the same time, despite its miraculous economic development, China remains a developing country, in which millions of people continue to live in poverty. These trends raise a variety of political and ethical questions about the expectations of China's role in international climate politics, including in relation to international justice and the allocation of responsibility.

Regarding theory, this book builds on and contributes to the English School of international relations, which maintains that states form an international society, the workings of which great powers have special responsibility to safeguard. Because I find it more interesting, as well as more important, to analyse how such responsibilities are constructed and allocated in practice, I assume that states – and individuals – have ethical responsibilities. Indeed, I argue that responsibility is always a situational ethics, the content of which is continuously made and remade via social practices in a process that I call *responsibilisation*. During that process, by using language and action, states and non-state actors attempt to create a common understanding of what it means to be responsible in international society in specific contexts. As a result, realising understandings of responsibility in international politics always involves competition. Today, when states define and distribute state responsibilities as well as great power responsibilities, the rising power of China undoubtedly plays a key role and will continue to do so in the future. In that sense, international climate politics is an especially interesting case of China's emerging notions of great power responsibility, for China has increasingly identified itself as a great power with great responsibility and, in turn, formulated ambitious climate policies to live up to that responsibility. As US leadership in great power responsibility for climate change declines in the era of President Donald Trump, China's emergence as a leader of global efforts to tackle climate change becomes more possible than ever before.

In an attempt to answer the looming question in international relations about how a great power should be defined in today's global era, I draw from the pluralism–solidarism debate within the English School. In particular, I focus on two international norms of responsibility – *great power responsibility* and *climate responsibility* – and investigate their interaction, as well as China's contribution to each. In that way, I demonstrate that responsibility is a principal criterion that states seeking recognition as great powers must fulfil and has constituted the ideological basis for the rule of the so-called 'great power club' since the early 1800s. With the end of the Cold War and China's rise in international status, the United States elevated responsibility as an imperative in the great power club. Accordingly, China's alleged irresponsibility can thus be viewed as the primary reason why it has not been accepted as a full member in the club. However, as climate responsibility increasingly becomes an international norm with which states, including great powers, must comply if they want to be and be recognised as responsible members of international society, China's central role in the institutionalisation of climate responsibility has become increasingly apparent. At the same time, the import of China's contribution cannot be understood without first understanding the political, economic and cultural-historical context in which China's practices of state responsibility have evolved. To that end, I study not only China's role in international climate negotiations but also the underlying domestic interests and values that have shaped its contributions to those negotiations.

Necessity of normative inquiry in international relations theory

Climate change is not only a neutral, natural and scientific phenomenon but, perhaps more importantly, a discursively created political problem that raises a range of moral questions about how humans should and do respond (e.g. Gardiner 2011; Gardiner et al. 2010). Instead of moral questions, however, traditional international relations studies have sought to discover in the first place whether and then, if appropriate, how and why states can cooperate to resolve global problems, including climate change. Realists argue that, in an anarchic world, there is little room for cooperation and always the risk of conflict. Conversely, liberals maintain that international cooperation is possible as well as necessary to address global challenges such as climate change and to prevent conflicts. By extension, many neorealist and neoliberal institutionalists focus on problem solving, especially regarding the potential role of international regimes in resolving conflicts and motivating cooperation among states. Although both sets of thinkers agree that states generally cooperate because 'it is in their interest to do so' (Hurrell & Kingsbury 1992, 23), they also tend to take actors and their interests as givens and pay little attention to the normative aspects of politics, including environmental politics. Unlike those ways of thinking, constructivism can offer unique insights into climate responsibility, for a significant, if not the most significant, part of international climate politics is 'discourse and dialogue concerning what policies or activities, ours as well as theirs, are desirable of advisable or appropriate or acceptable or tolerable or prudent or politic or judicious or justified in the circumstances' (Jackson 2000, 37). Climate change discourses define the nature of the phenomenon of climate change, its causes and its consequences and thus situate and control how climate-related issues are conceived and what actions are possible and prohibited in response. However, discourse is only one part of responsibility; the other, more critical part is its demonstration in action. Among its other limitations, constructivism does not consider the specific social contexts in which discourses and norms are produced. In particular, they tend to dismiss the intentionality of state behaviour as well as the role of (great) power in international society.

To clarify how states define and ought to define and fulfil their climate responsibility, in this book I integrate 'empirical knowledge and normative reasoning' (Reus–Smit 2013, 602). Although I draw inspiration from both liberal institutionalists' work on international organisations and constructivists' work on collective identities and discourses, my approach to climate responsibility differs profoundly from both. One reason for my departure from those ways of thinking is that they tend to frame environmental changes as technical and economic problems that have to be solved by collective inter-state action. Consequently, they fail to recognise that 'states are themselves (or alternatively, the state's system is itself, through generating certain practices on the part of states) prime environmental destroyers' (Paterson 2000, 2). Another more important reason is that liberal institutionalists and constructivists tend to treat norms and discourses as 'independent variables' and problem-solving endeavours such as international treaties and organisations as 'dependent variables' (Navari 2014, 209). In other words, they assume that norms and discourses cause change in a state's domestic and international behaviour via processes of socialisation (e.g. Finnemore & Sikkink 1998; Wendt 1999). Such approaches suggest that norms and practices exist 'out there' and that states 'internalise' them in their social interactions. By contrast, I emphasise that responsibilities are not given or static but always produced and reproduced in social interactions. Values and intentions are therefore important factors in how international responsibilities are defined, allocated and implemented by agents in specific contexts (cf. Navari 2018).

Among other reasons for regime theory proving an inadequate theoretical framework for studying climate responsibility is that the many important international treaties developed in recent decades have been unable to respond effectively to ecological challenges due to three major problems. First, international environmental agreements are compromises and do not provide an adequate basis for ending, preventing or even decreasing environmental degradation. In short, international regimes are too ineffectual to secure effective international environmental protection. States regularly avoid agreeing to legally binding obligations and instead prefer to commit to non-binding guidelines or principles because failing to meet such guidelines does not expose them to international criticism. On pressing issues such as climate change, despite decades of negotiations, states have failed to agree upon a sufficiently appropriate international treaty as well as to define their respective responsibilities. In particular, before the 2015 UN Conference on Climate Change in Paris, it seemed that international society was failing or had already failed to resolve the problem of climate change. Second, even when states manage to form international environmental agreements, their compliance is not guaranteed. That problem begs the question of how states can be ensured to implement and comply with the international rules that they have agreed to follow. When such actions are not taken, even the most serious international agreements become mere paper and fail to effect real difference. Third, international environmental agreements avoid reckoning with the 'question of why global environmental change occurs in the first place' (Paterson 2000, 3) and do not suggest that humans are part of Earth's ecosystems instead of separate from them. On the contrary, just as human practices have significant impacts on the environment, environmental changes have harmful impacts on human lives.

Given those shortcomings, this book builds on the English School theory, which is not only a theory about practices and norms, but a practice-guiding, normative theoretical framework that attempts to direct how human practices ought to be. In his landmark volume *The Anarchical Society* published in 1977, Hedley Bull coined the conception *international society*, which later emerged as a key concept of the English School. According to Bull (2002 [1977], 13), international society exists

> when a group of states, conscious of certain common interests and common values, form a society in the sense that they conceive themselves to be bound by a common set of rules in their relations with one another, and share in the working of common institutions.

The concept of international society lays the foundation for the normative framework of the English School: that states have rights and responsibilities due to their membership in international society. At a minimum, governments need to take the opinions and interests of others into consideration, and they cannot focus only on their narrowly defined national interests but are obliged to cooperate with others. For example, climate change politics does not support the normative logic of a sovereign state's right to do whatever it wants inside its borders, because states are bound to cooperate in order fulfil their climate- and environment-related obligations to other states. The capacity and willingness to accept and fulfil those responsibilities defines the status of their membership in international society, in which great powers have greater responsibilities than less powerful states. That idea makes the English School's theory unique in the field of international relations as the sole theoretical framework that stresses the special responsibilities of great powers. Other theoretical perspectives such as realism and neoliberalism, by contrast, focus on the balance of power or the sphere of interests of great powers but fail to consider the normative underpinnings of great power management. As Chris Brown (2004, 11) points out, to neorealists the idea of great powers' responsibilities towards international society as a whole does not make much sense because the idea of international society itself is unappealing to them. In the English School, however, great power management ranks among the common institutions of international society, in which great powers have a responsibility to sustain its efficient functioning. Nevertheless, environmental issues have been of interest to surprisingly few English School theorists, even though climate change can be conceived as a showcase for solidarist ethics (cf. Falkner 2012; Falkner & Buzan 2018; Hurrell 2007; Jackson 1996, 2000; Kopra 2018, forthcoming; Palmujoki 2013). Lately, however, some scholars in the English School have suggested that environmental stewardship is emerging or has already emerged as a new primary institution (Buzan 2004a, 186; Buzan 2014, 161–163; Falkner 2012, Falkner & Buzan 2018; Kopra forthcoming).

Institutional change: an English School approach

Change is a normal state of affairs in life, part of which in today's global era is international relations. In academic literature, change is usually explained by certain types of markers, including trends, great events and significant technological and social innovations (Holsti 2004, 7–12). A major technological development could, for example, dramatically lower the costs of mitigating climate change or carbon capture and storage and thus generate greater political will to shoulder broader climate responsibilities among states. In certain circumstances such changes can also produce new players on the international stage, including new sovereign states and non-state actors. In general, markers identify *when* change happens but do not specify *what kind* of change is happening (ibid., 12). In response, in *Taming the Sovereigns: Institutional Change in International Politics*, K. J. Holsti differentiates six types of concepts of change: change as novelty and replacement, change as addition or subtraction, change as increased or decreased complexity, change as transformation, change as reversion and change as obsolescence (ibid., 12–17). Nevertheless, those conceptualisations do not pinpoint *why* change occurs.

Many theorists of international relations, particularly neorealists and liberalist institutionalists, clarify change with reference to material factors. For example, Keohane and Nye (2012, 32–51) explain regime change in light of changes in economic and technological processes, overall power structures in the world, the power structure within specific contexts and power capabilities affected by international organisations. At times, change is the result of an external shock such as war, revolution or another crisis or shift in circumstances. For the time being, climate change has not caused a dramatic crisis; instead, its impacts have progressed slowly and thus remain invisible to general audiences, so to speak. If climate change were to cause a sudden humanitarian crisis, then states most likely take more urgent action. Not all changes in international society, however, can be explained by momentous events or material elements. For constructivists, the primary reason for change is the transformation of collective ideas.[2] They maintain that identities matter in inter-state relations and that when a state's identity changes, its behaviour in international society also changes accordingly. However, ideational change cannot alone explain institutional change because relationships informed by power and interests are important factors in shaping international society.

The English School underscores that both material and ideational factors induce change in international society. From that perspective, international institutions and practices are the most important markers and metrics of change in international society because they mirror international order and common ideas, problems, interests and norms among states during a given historical era (Holsti 2004, 18–19). Given their centrality in English School theory, it is thus surprising how premature agreement on the definitions, identity and role of institutions has been (Wilson 2012). Even Bull (2002 [1977]) did not elucidate what elements constitute a common institution, on what terms he chose his five common institutions or why he included others.[3] Recognising that shortcoming, in his seminal *From International to World Society? English School Theory and the Social Structure of Globalisation*, Barry Buzan (2004a, 171) highlighted the 'urgent need to acknowledge the centrality of primary institutions in English school theory, to generate consistency in the use and understanding of the concept and to make clear what does and does not count as a primary institution'.

Buzan's call has set in motion new institutionalists' theoretical debate about the (contemporary) primary institutions of international society and shunted secondary institutions 'into the realm of regime theory altogether' (Spandler 2015, 2; cf. Buzan 2004a, 163–167).[4] In fact, Buzan popularised the English School distinction of primary and secondary institutions in order to further elaborate upon Bull's common institutions of international society and how they organise that society. According to Buzan's (2004a, 181) definition, *primary institutions* are 'durable and recognised patterns of shared practices rooted in values held commonly by the members of interstate societies, amd [*sic*] embodying a mix of norms, rules and principles'. Although English School scholars do not agree upon what the primary institutions of international society are, they do agree that such institutions are critical to understanding inter-state relations because they, for instance, determine membership, organise relationships between states, facilitate coexistence and specify what legitimate international conduct is. Primary institutions are thus constitutive of international society and constantly shape processes in which responsibilities are made and remade. In this book, I study the ways in which the primary institution of great power management constrains and enables the institutionalisation of the international norm of climate responsibility. In methodological terms, I do not join the new institutionalist debate but follow the classical approach of the English School by studying how primary institutions shape the institutionalisation of norms in practice (cf. Jackson 2000; Jackson 2009).

Prior to Tonny Brems Knudsen's (2013) pre-theory of fundamental institutional change presented at the International Studies Association Annual Convention in 2013, English School theorists focused largely on the ways in which primary institutions induce change in international society. For example, Buzan (2004a, 186) contended that clashes among primary institutions are the 'key driving force' for institutional change in international society. However, Knudsen's paper pointed out that international organisations are central to the 'reproduction and working [of primary institutions], and therefore also to changes in their working' (Knudsen 2013, 18; cf. Knudsen 2018). In that sense, the relationship between primary and secondary institutions is not a one-way hierarchical relationship because they both shape each other. In explanation, Knudsen (2013, 16) identified two drivers of change: 'change in a fundamental institution' caused by 'changes in the practices by which the constitutive principles are reproduced or maintained' and the 'change of a fundamental institution', referring to 'changes in the constitutive principles themselves'. Notably, Knudsen (2013, 34) concluded that secondary institutions are the 'most important frameworks for the reproduction and change of fundamental institutions, and thus for the maintenance and development of international order and justice'. Moreover, his conceptualisation leaves room for the emergence of new primary institutions in the case that the constitutive principles of international society fundamentally change (Knudsen 2013, 17; Knudsen 2018).

Knudsen's ideas have inspired many English School theorists, including me, to study the role of secondary institutions in institutional change (e.g. Friedner Parrat 2014; Knudsen 2016; Kopra forthcoming; Navari 2016; Spandler 2015). In particular, they facilitated the emergence of an active working group led by Knudsen and Cornelia Navari within the International Studies Association's English School section. In a pioneering volume *International Organization in the Anarchical Society*, the group synthesises neoliberal institutionalist work on international institutions and classic English School theory on fundamental institutions of international society in response to a call for such synthesis issued by Robert O. Keohane (1989, 174) in the late 1980s. The volume clarifies that secondary institutions should be of interest to the English School not only because they provide material evidence of the existence of primary institutions but also given their unique role in promoting the change of international society and change within that society (cf. Knudsen & Navari 2018). Whereas neoliberal institutionalists study how the workings of international organisations could be improved as a means to solve global problems, the Navari-Knudsen working group studies how primary institutions shape the workings of secondary institutions and vice versa.

Arguably, the English School's theorisation on institutional change could benefit from incorporating ideas from neorealist and neoliberal institutionalists by paying closer attention to the role of state agency – or *statespeople*, to borrow Robert Jackson's (2000) term – and the domestic politics of great powers when investigating institutional change (cf. Navari 2018). The role of great powers is pivotal in the evolution of international norms and practices because powerful actors aim to define international rules in ways that serve their (domestic) interests and values (Clark 2011; Simpson 2004). Given such trends, this book focuses on not only China's international policies regarding the climate but also its historical development, domestic interests and social values that have shaped the state's international standing. Those trends also explain why secondary institutions are of special interest to me, for despite the impossibility of investigating an agent's influence on the evolution of a primary institution, a state's contribution to developing secondary institutions can be investigated in considerable depth. In addition to state agency, sub-national and non-state actors, including international organisations, non-governmental organisations, social and religious movements, scientists, media outlets, corporations, cities and provinces, influence the institutionalisation of secondary institutions in various ways and are thus significant subjects of change in international society (Clark 2007; Epstein 2008; Falkner 2012). In particular, they politicise new (environmental) problems, initiate or constrain international political agendas, produce and disseminate knowledge and participate in constructing the rules of international practices. Furthermore, they can influence the development of the domestic (climate) policies of individual states, as well as their positions in international (climate) negotiations.

At the same time, when investigating institutional change, we should not ignore the agency of individuals. After all, people influence and shape international practices, both negatively and positively (cf. Epstein 2008). For example, French leaders, especially Laurent Fabius, French foreign minister and president of the 2015 UN Conference on Climate Change in Paris, was widely commended for his role in the successful outcome of the conference. Or, had Hillary Clinton been elected as US president in 2016 instead of Donald Trump, the climate policy of the United States and thus of other states would have taken a drastically different path. At times, factors unrelated to a specific practice can nevertheless bear a significant impact upon that practice. For example, the election of George W. Bush as US president in 2000 likely did not relate to climate politics but influenced climate practices both locally and globally nonetheless. In China, the values, experiences and interests of the chair of the Chinese Communist Party undoubtedly exert significant influence on state practices due to the state's rather autocratic governance structure. In this book, however, I deliberately take a state-centric approach to climate responsibility, for two reasons. First, states continue to form the most important settings for negotiating practices at the international level, forming international treaties and putting them into practice at the local level. By contrast, the role of non-state actors remains quite limited; they can usually only influence and inspire the negotiating parties.[5] Second, the

Chinese Communist Party has shown very little interest in promoting the active participation of citizens in political decision-making processes at any level. In fact, the first Chinese environmental non-governmental organisation was founded two years after the establishment of the UN Framework Convention on Climate Change (UNFCCC), and their autonomy remains questionable. Given those deliberate dismissals, further research on the influence of non-state actors in both the evolution of international (climate) practices and China's notions of responsibility is highly recommended.

Secondary institutions in international society

Building upon the work of Knudsen, Navari, Keohane, Charlotta Friedner Parrat and Kilian Spandler, I define *secondary institutions* as 'stable, goal-oriented international bodies intentionally designed by international actors to manage and regulate common problems in specific areas of pragmatic issues and to govern cooperation according to collectively settled norms and rules, whether legally codified or not' (Kopra 2018). Secondary institutions include regimes, international organisations and even international rules that have emerged as established practices in the course of time (cf. Keohane 1989, 3–4.). Secondary institutions are always products of a time and are thus central to understanding power politics and the shared values of a particular era (cf. Navari 2018). However, their temporality does not mean that they are only 'arenas for acting out power relationships', as suggested by realists (Evans & Wilson 1992, 330). By contrast, secondary institutions are deliberately designed to solve global problems, and participants within them are usually willing to make concessions in order to identify workable solutions. In other words, participants not only engage in secondary institutions out of self-interest but also because they believe that participating is the right thing to do (Kopra 2018).

In this book, I aim to demonstrate that secondary institutions function as bridges between primary institutions and real-world politics performed by state and non-state actors on a daily basis. As I have argued elsewhere, the relationship between primary and secondary institutions is a reciprocal one (Kopra 2018; Kopra forthcoming). First, secondary institutions embed primary institutions in the quotidian workings of international relations. In general, I agree with Buzan (2004a) and Holsti (2004) that secondary institutions are empirical manifestations of primary ones. However, that perspective dismisses agency and interests in general and those of great powers in particular. Second, and by extension, secondary institutions also embody changes in the workings of the day-to-day international relations in primary institutions. The domestic practices of strong-minded, influential individual actors and especially power shifts in international relations can transform primary institutions via secondary institutions as well. For example, the global impacts of China's rise may not only transform everyday politics in secondary institutions but also gradually shape the constitutive principles of primary institutions. By way of secondary institutions, non-state actors can also shape existing primary institutions such as sovereignty or advance the emergence of new ones, as the cases of international environmental and human rights practices demonstrate. Secondary institutions therefore also function as bridges between international society and world society (Kopra 2018; Kopra forthcoming).[6]

In empirical terms, in this book I examine the institutionalisation of the international norm of climate responsibility, which cannot be located in a single secondary institution. On the contrary, members of international society can negotiate the definitions, allocation and implementation of climate responsibility within numerous international organisations. However, one special secondary institution does exist – namely, the UNFCCC – which plays a more central role in the construction of climate responsibility than any other international organisation by bridging the gap between the international norm of climate responsibility and real-life experience. The UNFCCC has no intrinsic value but is an instrumental regime that establishes a framework in which states and non-state agents can negotiate meanings, rules and appropriate choices of action to respond to climate change and to allocate climate responsibilities. As such, it formulates the infrastructure for participants to debate and enact their climate responsibility both globally and locally and provides a set of tools to do so. It also provides the infrastructure for derivative sub-practices in which its participants can engage, including the practices of climate finance and flexible market mechanisms. Furthermore, the UNFCCC facilitates the operationalisation of climate responsibility at national and local levels.

Consequently, observing the practices of the UNFCCC, including China's contribution to its processes, can provide valuable information about state and non-state actors' interpretations of climate responsibility. Given my interest in the historical evolution of climate responsibility, however, interviews and direct observations are not workable methods for my research. Moreover, it is exceedingly difficult to gain access to China's political circles in order to ask about their notions of responsibility. Therefore, I have relied upon textual analysis to read how both international climate practices and China's in particular have emerged and transformed. I trace the policies, treaties and statements by which states have negotiated, justified and agreed upon rules for international climate practices. My empirical corpus consists of four types of texts from 1968 to 2018: UN General Assembly resolutions and international agreements on the environment and climate, China's official policy documents such as white papers and other strategies, China's statements presented at UN climate change negotiations and both statements and acts of cooperation with other established and emerging great powers and international forums outside the UN system. I complement that corpus with newspaper articles, published academic studies and other relevant reports that offer information about how China has defined and performed its climate responsibility in real life.

State-centric solidarism and climate responsibility

This book contributes to the ongoing pluralist–solidarist debate among scholars in the English School about the possibility and potential of shared interests, norms, values, rules and institutions in international society (e.g. Bain 2014; Buzan 2014; Wilson et al. 2016). Pluralists regard states as the dominant actors in that society as well as emphasise the importance of state sovereignty. They focus chiefly on the 'is-side' of international ethics and ask what the practices of international society are and which norms organise and sustain international society and, in turn, notions of international responsibility. Pluralism's situational ethics raises questions about ways to manage collective problems that threaten the coexistence of states: how far states should go to put themselves at risk on the behalf of others, to what extent they have a moral obligation to rescue others such as victims of genocide, and to what extent they can ignore such humanitarian responsibilities if their national security and the lives of their citizens and soldiers is at risk. Regarding climate politics, an equivalent question asks to what extent states can promote their national (economic) interests at the expense of mitigating climate change and to what extent they have a moral duty to protect the climate. By contrast, solidarists assume that, in international society, states share a relatively high degree of norms, rules and institutions. Solidarism takes a cosmopolitan approach to the community of humankind – or *world society* in English School terms – and gives moral priority to the universal rights of individuals over state sovereignty. In other words, solidarism considers humankind as a moral referent and raises questions about (humanitarian) justice. In practice, the English School remains 'sharply divided over the extent to which solidarism remains premature' (Linklater & Suganami 2006, 229), and 'few if any' English School theorists have suggested a cosmopolitan world society without states as 'either a theoretical or practical option', though many have doubted the potential for states to transcend pluralism (Buzan 2014, 119). Nonetheless, solidarism has served as the 'key source' of normative discussions characteristic of the English School (ibid.,118). At the heart of the debate have been human rights, especially pertaining to the question of humanitarian intervention and the responsibility to protect (e.g. Wheeler 2000).

Because climate change is an ethical global problem I find the concept of responsibility particularly apt for scrutinising international practices in response to climate change. Accordingly, I have deliberately chosen to study the norm of climate responsibility, which problematises the power relations that shape international climate practices as well as emphasises the finality and future orientation of those practices, for whatever states decide to do today affects the wellbeing of future generations, and it will be incredibly difficult, if not impossible, to remedy today's wrongheaded or ethically irresponsible policy choices in the future. My conceptual choice also underscores that the terms *responsibility* and *duty* are not synonyms, although they are sometimes used interchangeably. Whereas some people would argue that *duty* better describes the moral agency and obligations of states, I choose to continue the conceptual tradition of both the classical literature of the English School and international climate discourse. Moreover, because the concept of responsibility highlights the significance of good outcomes, it better suits international climate politics, the outcomes of which are more important than the performance of certain actions.

I formulate my approach to responsibility by combining elements from the works of the English School scholars, mainly Jackson and Buzan, with the framework of ecocentric thinking, as in the work of Robyn Eckersley. Both Buzan and Eckersley argue that when we discuss ethics, we should not focus on the polarisation of mutually exclusive positions such as realism–liberalism, pluralism–solidarism or anthropocentrism–naturocentrism but instead consider moral standings as shifting positions on a broad spectrum of moral orientations. Buzan (2004a, 139) suggests that the pluralist and solidarist perspectives of the English School could be reconstructed 'not as mutually exclusive positions, but as positions on a spectrum representing, respectively, thin and thick sets of shared norms, rules and institutions'. If seen as the ends of a spectrum, then those perspectives would strengthen 'the position of international society as the via media between state-centric realism and cosmopolitan world society' (ibid., 50). Likewise, Eckersley (1992, 35–47) points out that the contemporary division of anthropocentrism and naturocentrism showcases the 'opposing poles of a wide spectrum of differing orientations toward nature' and that most recent studies in environmental philosophy fall 'between these two poles'. In order to blend and mix the two camps, Buzan (2014, 114–118) introduces the concept of state-centric solidarism, which I choose to adopt in this book. Ontologically, state-centric solidarism is similar to pluralism because it recognises that current international society is state-centric and that the potential for world society thus remains limited. Politically and morally, however, state-centric solidarism more closely aligns with solidarism because it acknowledges humankind as the moral referent object. State-centric solidarism is not only about world order and co-existence but also about cooperation and the pursuit of collective objectives. In Buzan's (2014, 116) words, it is about the 'possibility that states can collectively reach beyond a logic of coexistence to construct international societies with a relatively high degree of shared norms, rules and institutions among them'.

Great powers in international society

Theorists in international relations have debated the definitions and roles of great powers in international politics since the mid-eighteenth century because, as Kenneth Waltz (1979, 72) puts it, 'In international politics, as in any self-help system, the units of greatest capability set the scene of action for others as well as for themselves.' Realists define *national power* and therefore *great powers* mainly in terms of material capabilities but do not entirely ignore social capabilities.[7] Many realists find it useful to explain changes in international systems by counting the number of great powers and analysing the shifting distribution of power among states (cf. Waltz 1979, 131). English School scholars, who are interested in historical developments in international relations, generally agree; instead of defining what constitutes a great power, they maintain that 'it is easier to answer it [the question of defining *great power*] historically, by enumerating the great powers at any date' (Wight 1999 [1946], 41). As a result, many scholars within and outside the English School have offered lists of previous, contemporary and potential future great powers.[8] Such lists are, however, often incoherent, and even Wight gives inconsistent reasons why he ranked some states as great powers (Buzan 2004b, 59). By contrast, constructivists emphasise the role of identities and social interaction when identifying great powers. According to Brittingham (2007, 84), a 'great power is an identity that must be enacted by a state, and recognised and reinforced by its peers'. In addition to those general definitions of *great powers*, some scholars classify special categories of great powers – namely, superpowers and regional powers – based on the 'operational range of power holds' (Buzan 2004b, 50–53).

In general, the English School conceptualisation of great power integrates realist and constructivist perspectives. Its perspective holds that though great powers need to have certain material capabilities, the status of great power is above all an identity created in interaction with other states. For the English School, power is a 'social attribute' that must be placed 'side by side with other quintessentially social concepts such as prestige, authority, and legitimacy' (Hurrell 2007, 39). Instead of providing a clear-cut definition of *great power*,[9] thinkers in the English School describe at least five important dimensions of what a state must fulfil in order to be and be recognised as a great power (e.g. Bull 2002 [1977]; Buzan 2004b; Cui & Buzan 2016; Hurrell 2007; Jackson 2000; Jones 2014; Simpson 2004; Wheeler 2000; Wight 1999 [1946]). First, great powers have to have a certain level of capabilities. In Bull's (2002 [1977]) opinion, it is essential that great powers rank their military strength as being superior to that of other states. In the post-Cold War era, however, military strength has become a less important dimension of great power, whereas the significance of soft power and credibility has increased. Second, great powers can exist only in a plurality. In mainstream international relations theory, the international society or system is anarchical; hence, two or more great powers are always necessary. Realists would explain that situation as stemming from the balancing role of great powers in the international system, whereas liberalists regard multilateral organisations as having an important role in the prevention of inter-state conflicts. As constructivists and the English School would explain, by contrast, great power status is based on membership in a social group with a shared identity and is thus always the result of ranking the comparable statuses of states. 'When we speak of great powers', Bull (2002 [1977], 194) writes, 'we imply ... the existence of a club with a rule of membership'. Likewise, according to Wight (1999 [1946], 42), great powers have a 'tendency to club together as a kind of directorate and impose their will on the states-system'.

Third, being great power means having a social identity that shapes how certain states perceive themselves as well as how others treat them. Nationalism thus matters in power politics; to some extent, great powers are great because their citizens consider or wish that their country is greater than other states. Like individuals, states construct their identities in social interaction and define themselves in relation to others. To quote Buzan (2004b, 61), 'great power identity (or indeed any international identity) is a reciprocal construction composed of the interplay between a state's view of itself and the view of it held by the other members of international society'. Fourth, even if a state reaches a certain level of material capacity and has a certain national identity, it does not automatically become a great power but has to be recognised and accepted as such by others. Thus, it is important to 'distinguish between power that is based on relations of domination and force, and power that is legitimate because it is predicated on shared norms' (Wheeler 2000, 2). Given the Eurocentrist nature of contemporary international society, it is usually the West whose recognition matters the most. Consequently, China's friendship with rogue states such as North Korea does not raise its international status, for its great power status must be recognised by the United States and the European Union.

Last – and most importantly from the perspective of this book – great powers have internally and externally recognised rights and responsibilities. In contrast to other states, great powers have the capability and legally authorised right to 'play a part in determining issues that affect the peace and security' of international society (Bull 2002 [1977], 196). That right comes with the responsibility to modify their 'policies in the light of the managerial responsibilities they bear' (ibid.). International responsibility is assumed to be more or less causal; the greater the power of a state, the greater the international effects its domestic and foreign policy will have and the greater its responsibility for the collective wellbeing of international society.[10] Within the English School, special responsibilities have thus been largely attached to great powers, which have 'fundamental global capabilities and responsibilities that minor or medium powers do not' (Jackson 2000, 21). That normative contribution of the English School adds an important question to discussions of international climate responsibility – namely, whether is it justified to assume that great powers should shoulder more responsibility for mitigating climate change than other, less powerful states. That question sparks an important discussion that I seek to engage with in this book.

Organisation of the book

This book examines the ways in which the primary institution of great power management has shaped international climate negotiations and China's role in those processes following its entrance into the great power club. By analysing China's contribution to the institutionalisation of the international norm of climate responsibility in the UNFCCC, it contributes to the Navari-Knudsen working group's research agenda on the role of secondary institutions in international society. At the same time, because climate responsibility does not materialise from the UNFCCC but needs to be implemented as national policies and practised at the grass-roots level, the book also scrutinises how China defines and enacts its climate responsibility at the domestic level.

Following this introduction, chapter 2 briefly introduces a general framework of state responsibility in international society and distinguishes the general responsibilities of all states from the special responsibilities of great powers. The latter is the book's chief focus, for the basic premise of the English School is that great powers have a special responsibility to sustain and organise international society.

Chapter 3 investigates China's notions of responsibility by questioning what the concept of responsibility has meant to different generations of Chinese leadership, as well as for what and to whom the Chinese government considers itself to be responsible and why. In so doing it discusses the necessary political and historical contexts, as well as the underlying interests and values, that have shaped contemporary China's conceptualisations of responsibility.

Chapter 4 examines the particular notion of great power responsibility in the context of international climate politics. To that end it sketches an ethical framework of great power responsibility for climate change and debates the extent to which such responsibility has been acknowledged and acted upon within international society. In more empirical terms, the chapter discusses the requirements a state needs to fulfil to become an accepted member of the great power club, especially according to the notions of great power responsibility of the UN Security Council, China and the United States. By probing associated notions of great power responsibility in international climate politics and other fields, the chapter traces China's evolving identity as a great power in particular depth.

Chapter 5 discusses the process of institutionalising the norm of climate responsibility and China's contribution to that process. It begins by reviewing the environmental awakening of international society, particularly by studying how major international conferences addressing the environment and climate have attempted to articulate states' environmental responsibilities and motivate their fulfilment of those responsibilities. Thereafter, the chapter examines how general and special responsibilities have been defined and assigned by the UN climate regime and pays close attention to how China's changing notions of responsibility have shaped its role in international climate negotiations.

Chapter 6 investigates how China has demonstrated its climate responsibility in actions at the international and local levels. Since the UN climate regime does not specify how states have to implement their climate responsibility, it falls to each state to decide which measures it will implement to fulfil its responsibility. The chapter especially focuses on the sorts of policies and actions that China has chosen to realise its climate responsibility and reviews the key drivers of its climate practices.

Last, chapter 7 summarises the chief contributions of the book and discusses prospects for climate responsibility in the coming years. It also makes a few recommendations regarding how international climate practices can or even should be transformed in order to strengthen climate responsibility and enhance efforts to mitigate climate change in the future.

Notes

1 I discuss these issues elsewhere as well (cf. Kopra 2018; Kopra forthcoming).
2 See Legro (2005) for a detailed framework of how ideas influence continuity and change in international society.
3 For enlightened guesses of Bull's reasons, see Buzan (2014, 97–98) and Schouenborg (2014, 80–81).
4 As Wilson (2012, 580) observes, new institutionalists such as Buzan and Schouenborg methodologically depart from the English School's traditional focus on social reality and seek to construct abstract analytical categories instead.
5 However, see Epstein (2008) for a rare case in which international practice has emerged and diffused via a bottom-up process induced by non-state actors.
6 On the relationship between international society and world society, see Buzan (2004a), Clark (2007) and Williams (2014).
7 According to Morgenthau (1993, ch. 9), the components of national power include geography, natural resources, industrial capacity, military preparedness, population, national character, national morale, the quality of diplomacy and the quality of government. For Waltz (1979, 131), a state's power ranking depends on its capabilities: the size of its population and territory, its resource endowment, its economic capability, its military strength and its political stability and competence.
8 Scholars such as Kennedy (1988) include only states that meet Western (material) definitions of *great powerhood* on their lists of great powers. By contrast, others recognise the role of non-Western states in the history of great power politics. Among them, Black (2008) addresses China's changing role in great power politics from 1500 to the present day.
9 However, Simpson's (2004, 68) definition of *legalised hegemony* suggests a useful definition of *great power* as well: 'the existence within an international society of a powerful elite of states whose superior status is recognised by minor powers as a political fact giving rise to the existence of certain constitutional privileges, rights and duties and whose relations with each other are defined by adherence to a rough principle of sovereign equality'.
10 This is only a general correlation. In practice, a state's global influence also depends on its traditions, image, identity, experience and know-how, among other things. Small states such as Switzerland and Scandinavian countries may play an important diplomatic role in the resolution of a conflict or the formulation of international norms, for instance.

References

Bain, William. 2014. 'The pluralist–solidarist debate in the English School'. In Cornelia Navari & Daniel M. Green (eds), *Guide to the English School in International Studies.* Chichester: John Wiley & Sons, 159–169.

Black, Jeremy. 2008. *Great Powers and the Quest for Hegemony: The World Order since 1500*. London: Routledge.

Brittingham, Michael Alan. 2007. 'China's contested rise: Sino-US relations and the social construction of great power status'. In Sujian Guo & Shiping Hua (eds), *New Dimensions of Chinese Foreign Policy.* Plymouth: Lexington Books, 83–108.

Brown, Chris. 2004. 'Do great powers have great responsibilities? Great powers and moral agency'. *Global Society* 18:1, 5–19.

Bull, Hedley. 2002 [1977]. *The Anarchical Society: A Study of Order in World Politics*, 3rd edition. Basingstoke: MacMillan Press.

Buzan, Barry. 2014. *An Introduction to the English School of International Relations.* Cambridge: Polity Press.

Buzan, Barry. 2004a. *From International to World Society? English School Theory and the Social Structure of Globalisation.* Cambridge: Cambridge University Press.

Buzan, Barry. 2004b. *The United States and the Great Powers: World Politics in the Twenty-First Century.* Cambridge: Polity Press.

Chan, Gerald. 2006. *China's Compliance in Global Affairs: Trade, Arms Control, Environmental Protection, Human Rights.* Singapore: World Scientific Publishing.

Chen, Zhimin. 2009. 'International responsibility and China's foreign policy'. In Masafumi Iida, *China's Shift: Global Strategy of the Rising Power.* NIDS Joint Research Series no 3. Tokyo: National Institute for Defense Studies, 7–28.

Chen, Gang. 2012. *China's Climate Policy.* New York: Routledge.

Chen, Gang. 2009. *Politics of China's Environmental Protection: Problems and Progress.* Singapore: World Scientific Publishing.

Clark, Ian. 2014. 'International society and China: The power of norms and the norms of power'. *Chinese Journal of International Politics* 7:3, 315–340.

Clark, Ian. 2011. *Hegemony in International Society.* New York: Oxford University Press.

Clark, Ian. 2007. *International Legitimacy and World Society.* New York: Oxford University Press.

Cui, Shunji & Barry Buzan. 2016. 'Great power management in international society'. *Chinese Journal of International Politics* 9:2, 181–210.

Deng, Yong. 2008. *China's Struggle for Status: The Realignment of International Relations.* Cambridge: Cambridge University Press.

Eckersley, Robyn, 1992. *Environmentalism and Political Theory: Toward an Ecocentric Approach*. London: UCL Press.

Ella, Doron. 2016. 'China and the United Nations framework convention on climate change: The politics of institutional categorization'. *International Relations of the Asia-Pacific* 17:2, 233–264.

Epstein, Charlotte. 2008. *The Power of Words in International Relations.* Cambridge: MIT Press.

Evans, Tony & Peter Wilson. 1992. 'Regime theory and the English School of international relations: A comparison'. *Millennium: Journal of International Studies* 21:3, 329–351.

Falkner, Robert. 2012. 'Global environmentalism and the greening of international society'. *International Affairs* 88:3, 503–522.

Falkner, Robert & Buzan, Barry. 2018. 'The emergence of environmental stewardship as a primary institution of global international society'. *European Journal of International Relations.*

Finnemore, Martha and Kathryn Sikkink. 1998. 'International norm dynamics and political change'. *Intemational Organization* 52:4, 887–917.

Friedner Parrat, Charlotta. 2014. 'International organization in international society: UN reform from an English School perspective'. *Journal of International Organization Studies* 5:1, 7–21.

Foot, Rosemary. 2001. 'Chinese power and the idea of a responsible state'. *The China Journal* 45, 1–9.

Gardiner, Stephen M. 2011. *A Perfect Moral Storm: The Ethical Tragedy of Climate Change.* Oxford: Oxford University Press.

Gardiner, Stephen M., Simon Caney, Dale Jamieson & Henry Shue (eds) 2010. *Climate Ethics: Essential Readings.* Oxford: Oxford University Press.

Gill, Bates. 2007. 'China becoming a responsible stakeholder'. *Carnegie Endowment for International Peace,* June 11. Accessed 20 February 2017. http://carnegieendowment.org/files/Bates_paper.pdf.

Gill, Bates, Dan Blumenthal, Michael D. Swaine & Jessica Tuchman Mathews. 2007. 'China as a responsible stakeholder'. *Carnegie Endowment for International Peace,* June 11. Accessed 20 February 2017. www.carnegieendowment.org/2007/06/11/china-as-responsiblestakeholder/2kt.

Gong, Gloria Jean. 2011. 'What China wants: China's climate change priorities in a post-Copenhagen world'. *Global Change, Peace & Security* 23:2, 159–175.

Harris, Paul G (ed.) 2011. *China's Responsibility for Climate Change: Ethics, Fairness and Environmental Policy.* Bristol: Policy Press.

Harris, Paul G. & Hongyuan Yu. 2009. 'Climate change in Chinese foreign policy: internal and external responses'. In Harris, Paul G. (ed.), *Climate Change and Foreign Policy.* New York: Routledge, 53–67.

Holsti, K. J. 2004. *Taming the Sovereigns: Institutional Change in International Politics.* Cambridge: Cambridge University Press.

Hurrell, Andrew. 2007. *On Global Order: Power, Values, and the Constitution of International Society.* Oxford: Oxford University Press.

Hurrell, Andrew & Benedict Kingsbury. 1992. 'The international politics of the environment: An introduction'. In Andrew Hurrell & Benedict Kingsbury (eds), *The International Politics of the Environment.* New York: Oxford University Press, 1–47.

Jackson, Robert H. 2009. 'International Relations as a Craft Discipline'. In Cornelia Navari (ed.), *Theorising International Society: English School Methods.* New York: Palgrave MacMillan, 21–38.

Jackson, Robert H. 2000. *The Global Covenant.* Oxford: Oxford University Press.

Jackson, Robert H. (1996): 'Can international society be green?'. In Rick Fawn & Jeremy Larkins (eds), *International Society after the Cold War: Anarchy and Order Reconsidered.* Basingstoke: MacMillan Press, 172–192.

Jin, Canrong. 2011. *Big Power's Responsibility: China's Perspective.* Translated by Tu Xiliang. Beijing: China Renmin University Press.

Jones, Catherine. 2014. 'Constructing great powers: China's status in a socially constructed plurality'. *International Politics* 51:5, 597–618.

Kennedy, Paul. 1988. *The Rise and Fall of the Great Powers: Economic Change and Military Conflict from 1500 to 2000*. London: Unwin Hyman.
Keohane, Robert Owen. 1989. *International Institutions and State Power: Essays in International Relations Theory*. Boulder, CO: Westview.
Keohane, Robert O. & Joseph S. Nye, Jr. 2012. *Power and Interdependence*, 4th edition. Boston: Longman.
Knudsen, Tonny Brems. 2018. 'Fundamental institutions and international organizations: Theorizing continuity and change'. In Tonny Brems Knudsen & Cornelia Navari (eds), *International Organization in the Anarchical Society: The Institutional Structure of World Order*. New York: Palgrave Macmillan.
Knudsen, Tonny Brems. 2016. 'Solidarism, pluralism and fundamental institutional change'. *Cooperation and Conflict* 51:1, 102–109.
Knudsen, Tonny Brems. 2013. 'Master institutions of international society: Theorizing continuity and change'. 8th Pan-European Conference on International Relations, European International Studies Association, Warsaw.
Knudsen, Tonny Brems & Cornelia Navari (eds).2018. *International Organization in the Anarchical Society: The Institutional Structure of World Order*. New York: Palgrave Macmillan.
Kopra, Sanna. Forthcoming. 'China and the UN climate regime: Climate responsibility from an English School perspective'. *Journal of International Organizations Studies.*
Kopra, Sanna. 2018. 'China, Great Power Management, and Climate Change: Negotiating Great Power Climate Responsibility in the UN'. In Tonny Brems Knudsen & Cornelia Navari (eds), *International Organization in the Anarchical Society: The Institutional Structure of World Order*. New York: Palgrave Macmillan.
Morgenthau, Hans J. 1993. *Politics Among Nations: The Struggle for Power and Peace*. Brief edition revised by Kenneth W. Thompson. Boston: McGraw-Hill.
Navari, Cornelia. 2018. 'Modelling the relations of fundamental institutions and international organizations'. In Tonny Brems Knudsen & Cornelia Navari (eds), *International Organization in the Anarchical Society: The Institutional Structure of World Order*. New York: Palgrave Macmillan.
Navari, Cornelia. 2016. 'Primary and secondary institutions: Quo vadit?'. *Cooperation and Conflict* 51:1, 121–127.
Navari, Cornelia. 2014. 'English School methodology'. In Cornelia Navari & Daniel M. Green (eds), *Guide to the English School in International Studies*. Chichester: John Wiley & Sons, 205–221.
Legro, Jeffrey W. 2005. *Rethinking the World: Great Power Strategies and International Order*. Ithaca, NY: Cornell University Press.
Linklater, Andrew & Hidemi Suganami. 2006. *The English School of International Relations: A Contemporary Reassessment*. New York: Cambridge University Press.
Lynas, Mark. 2009. 'How do I know China wrecked the Copenhagen deal? I was in the room'. *The Guardian*, 22 December. Accessed 26 February 2018. www.theguardian.com/environment/2009/dec/22/copenhagen-climate-change-mark-lynas.
Marks, Danny. 2010. 'China's climate change policy process: Improved but still weak and fragmented'. *Journal of Contemporary China* 19:67, 971–986.
Moore, Scott. 2011. 'Strategic imperative? Reading China's climate policy in terms of core interests'. *Global Change, Peace & Security* 23:2, 147–157.
Palmujoki, Eero. 2013. 'Fragmentation and diversification of climate change governance in international society'. *International Relations* 27:2, 180–201.

Paterson, Matthew. 2000. *Understanding Global Environmental Politics: Domination, Accumulation, Resistance.* Basingstoke: Palgrave.

Patrick, Stewart. 2010. 'Irresponsible stakeholders? The difficulty of integrating rising powers'. *Foreign Affairs* 89:6, 44–53.

Porter, Andrew. 2009. 'China and America to blame for Copenhagen failure, says Brown'. *The Telegraph*, 21 December. Accessed 26 February 2018. http://www.telegraph.co.uk/news/politics/6859567/Gordon-Brown-Copenhagen-China.html.

Reus-Smit, Christian. 2013. 'Beyond metatheory?'. *European Journal of International Relations* 19:3, 589–608.

Schouenborg, Laust. 2014. 'The English School and institutions: British institutionalists?'. In Cornelia Navari & Daniel M. Green (eds), *Guide to the English School in International Studies.* Chichester: John Wiley & Sons, 77–89.

Scott, David. 2010. 'China and the "responsibilities" of a "responsible" power: The uncertainties of appropriate power rise language'. *Asia-Pacific Review* 17:1, 72–96.

Simpson, Gerry. 2004. *Great Powers and Outlaw States: Unequal Sovereigns in the International Legal Order.* Cambridge: Cambridge University Press.

Shambaugh, David. 2013: *China Goes Global: The Partial Power.* New York: Oxford University Press.

Spandler, Kilian. 2015. 'The political international society: Change in primary and secondary institutions.' *Review of International Studies* 4:3, 601–622.

Vidal, John. 2009. 'Ed Miliband: China tried to hijack Copenhagen climate deal'. *The Guardian.* December 20. Accessed 26 February 2018. www.theguardian.com/environment/2009/dec/20/ed-miliband-china-copenhagen-summit.

Waltz, Kenneth N. 1979. *Theory of International Politics.* New York: McGraw-Hill.

Wendt, Alexander. 1999. *Social Theory of International Politics.* Cambridge: Cambridge University Press.

Wheeler, Nicholas J. 2000. *Saving Strangers: Humanitarian Intervention in International Society.* Oxford: Oxford University Press.

Wight, Martin. 1999 [1946]. *Power Politics.* Edited by Hedley Bull and Carsten Holbraad. London: Leicester University Press.

Wilson, Peter. 2012. 'The English School meets the Chicago School: The case for a grounded theory of international institutions'. *International Studies Review* 14, 567–590.

Williams, John. 2014. 'The international society–world society distinction'. In Cornelia Navari & Daniel M. Green (eds), *Guide to the English School in International Studies.* Chichester: John Wiley & Sons,127-142.

Wilson, Peter, Yongjin Zhang, Tonny Brems Knudsen, Paul Sharp, Cornelia Navari & Barry Buzan. 2016. 'The English School in retrospect and prospect: Barry Buzan's An Introduction to the English School of International Relations: The Societal Approach'. *Cooperation and Conflict* 51:1, 94–136.

Yeophantong, Pichamon. 2013. 'Governing the world: China's evolving conceptions of responsibility'. *Chinese Journal of International Politics* 6:4, 329–364.

Xia, Liping. 2001. 'China: A responsible great power'. *Journal of Contemporary China* 10:26, 17–25.

Zhao, Suisheng. 2013. 'Core interests and great power responsibilities: The evolving pattern of China's foreign policy'. In Xiaoming Huang & Robert G. Patman (eds), *China and the International System: Becoming a World Power.* New York: Routledge, 32–56.

Zhang, Yongjin & Greg Austin (eds). 2001. *Power and Responsibility in Chinese Foreign Policy.* Canberra: ANU E Press.

2 Responsibility in international society

Within and outside the English School it is widely accepted that organisations such as states, corporations and institutions are moral agents. After all, they are human constellations, and humans cannot evade moral questions of right and wrong (cf. Erskine 2003; French 1984; French & Wettstein 2006; Mayer & Vogt 2006). States – more precisely, the legitimate representatives of states – bear ultimate responsibility for peaceful coexistence in international society because they have the highest authority to make decisions and take actions, including ones concerning the use of coercive power in their respective sovereign territories. This chapter offers an approach to responsibility informed by the English School by discussing the ways in which responsibilities are negotiated, allocated and implemented in international society. I begin with an overview of theoretical accounts of responsibility by asking what *responsibility* means in legal and moral terms. I deliberately avoid discussing rights at length given the extensive body of literature on human rights, including environmental ones.[1] Clearly, responsibilities are tied to rights; if someone has a right, then others have, at a bare minimum, a corresponding responsibility to refrain from infringing that right. Next, I argue that responsibilities are always constructed in social processes that I refer to collectively as *responsibilisation*. As members of international society, states have to participate in fulfilling and assigning responsibilities within that society. In the last section, I examine how the practices of state responsibility materialise in real life and ask what sort of multidimensional responsibilities states bear and ought to bear regarding climate change. To that end, the English School provides insightful standpoints for investigating those questions.

Many meanings of *responsibility*

Responsibility is a nebulous concept, and talking about what it means to be responsible warrants the consideration of numerous dimensions. For instance, we have to clarify the subject and the object of responsibility: who or what is responsible for what, and to whom is the subject accountable? By extension, we also have to draw a distinction between 'identifying responsibility and assigning it' (Miller 2007, 84). According to David Miller, the former task concerns determining 'who, if anybody, meets the relevant conditions for being responsible', whereas the latter 'involves a decision to attach certain costs or benefits to an agent, whether or not the relevant conditions are fulfilled' (ibid.). In his landmark book *Punishment and Responsibility*, H. L. A. Hart (1968, 212–230) characterises four types of responsibility: role-responsibility, causal-responsibility, liability-responsibility and capacity-responsibility. *Role-responsibility* suggests that all social roles have their own 'sphere of responsibility'; each position in a social organisation is attached to short-term tasks or duties whose fulfilment somehow advances the goals of the organisation. Accordingly, a responsible person is someone who takes those duties seriously and behaves correspondingly (ibid., 212–213). By contrast, *causal-responsibility* describes the relationship between cause and outcome; for instance, 'A is responsible for Y' means that Y is a direct or indirect result of what A has done. As Hart's (1968, 211) story about a drunken captain demonstrates, it is also possible for things, conditions and events to be responsible for results. In that sense, no moral blame is attached to causal responsibility. To some extent, causal responsibility is assumed to be an important, although not an entirely sufficient, precondition for moral and legal responsibility, or what Hart calls *liability-responsibility*. Hart's conceptualisation of liability-responsibility distinguishes legal from moral liability-responsibility. When considering liability in the context of legal responsibility, 'A is responsible for Y' means that A is somehow at fault in causing Y and can be rightfully punished for it in legal terms. For instance, a person who breaks the law is usually regarded to be liable if a certain range of necessary and sufficient (e.g. psychological) conditions are met.[2] However, when considering moral liability-responsibility, 'A is responsible for Y' means that A is blameworthy for Y, which can be rightfully disapproved in moral terms. Similar to legal liability, moral liability-responsibility also presupposes that a person has certain normal capacities, including freedom of choice. Last, *capacity-responsibility* can be understood from the expression 'A is responsible for his/her actions' if he or she possesses a normal (e.g. psychological) capacity of understanding and control (Hart 1968, 227). In international justice, capacity is an important precondition for judging a state's responsibilities. If a state has no capacity to act appropriately, then how can it be held responsible?

Many scholars distinguish legal from moral responsibility. At first glance, both types presumably refer to the same sort of responsibility, at least in an ideal world. However, mostly for practical reasons, they are not always identical. The greatest difference between them is that legal responsibility is always judged by a jurisdiction, whereas moral responsibility is assessed by morals, a 'kind of internal law, governing those inner thoughts and volitions which are completely subject to the agent's control, and administered before the tribunal of conscience' (Feinberg 1970, 33). Another distinguishing feature of legal compared to moral responsibility is their temporal orientation. The focus of legal responsibility is always retrospective; for instance, a court asks whether A is guilty of doing harm to X. Accordingly, a person cannot be held legally responsible for something that he or she has not done (or failed to do). Usually, assessing moral responsibility follows a similar logic; we cast moral blame upon someone for something that he or she has done or failed to do. Moral responsibility, however, can also be prospective, as the concept of sustainable development demonstrates. Often, both legal responsibility and moral responsibility contain a causal component; one is either legally or morally at fault for the harm that one has committed.[3] Thus, a person cannot be held responsible for something that has not happened because of their actions. With respect to climate change, however, it is exceptionally difficult, if not impossible, to identify a single state or private enterprise as being at fault for causing climate change. On its own, causality is not a justified factor of responsibility; however, there are other conditions of moral and legal judgement as well, including intentions, motives and choices, which are important when determining responsibility. A person cannot be held responsible for an event that occurs incidentally or due to factors beyond their control. To be morally blameworthy, a person is usually expected to have had an opportunity and the freedom to 'have acted otherwise than he did' (Ross 1975, 15). Thus, free will and the absence of coercion is an important condition; a person has to have acted voluntarily in order to be held responsible for an event (May 1992, 16). However, that dynamic does not necessarily mean that people are held responsible for their actions only: omissions matter as well.

Consequentialist theorists, who emphasise the significance of beneficial outcomes, claim that a person can also be held morally responsible for his or her omissions. In fact, they make no distinction between consequences resulting from acts or omissions. That view opposes deontologism, whose proponents seek to determine why agents do what they do and thus ask what the real motives are behind their actions. 'What matters much more to them [deontologists]', Goodin (1995, 47) writes, 'are individuals' [or states'] motives and intentions. They also insist that it be done, and be seen to be done, for the right reasons'. Because deontologists pay less attention to the consequences of acts, they do not hold persons responsible for their omissions. The distinction between positive and negative responsibility is often demonstrated in terms of the difference between killing (i.e. in which an agent plays an active role) and letting die (i.e. in which an agent plays a passive role), which have the same consequence (Vanderheiden 2008, 151–152). However, in climate politics, it does not matter what an agent's motives are in taking action; climate mitigation does not have to be performed for humanitarian, environmental or other altruistic reasons but could be a side-effect of energy security projects or the development of so-called 'green jobs'.[4] What counts is that states shoulder their responsibility to mitigate climate change and cut their greenhouse gas emissions, to which end they can choose the most suitable and cost-efficient mitigating actions. Although some states prefer to rely upon market-oriented economic mechanisms, some choose to establish new regulations and taxes and some pursue new technologies, all such means seek to achieve the goal of climate protection.

Feinberg (1970, 31–32) points out that moral responsibility 'cannot be a matter of luck', as it often is in the law, but 'must be something one can neither escape by good luck nor tumble into through bad luck'. Feinberg illustrates that concept – what Thomas Nagel (1979, 24–38) calls 'moral luck' – with the following example:

> One man shoots another and kills him, and the law holds him responsible for the death and hangs him. Another man, with exactly the same motives and intentions, takes careful aim and shoots at his enemy but misses because of a last-minute movement of his prey or because of his own bad eyesight. The law cannot hold him responsible for a death because he has not caused one; but, from the moral point of view, he is only luckier than the hanged murderer.
>
> (Feinberg 1970, 31–32)

When it comes to international climate politics, Russia has benefitted from the decision that the benchmark year of the Kyoto Protocol is 1990, a year prior to the collapse of the Soviet Union and, in turn, the closure of many inefficient factories. Because post-Soviet Russia's emissions are therefore compared to the emissions of all 15 former Soviet republics, the country does not need to not do anything in order to comply with targets for reducing international emissions. In that case, Feinberg's example, when applied analogously, suggests that international climate law cannot hold Russia irresponsible even if it does nothing to mitigate climate change; however, from a moral standpoint, Russia is simply luckier than other polluting states. By contrast, Iceland is one of the few states operating entirely on renewable energy, which, though outstanding, is not the result of political decision making. More accurately, Iceland, as a small volcanic island blessed with geothermal energy sources, is simply a very lucky country in terms of renewable energy resources. To some extent, both Russia's and Iceland's fulfilment of climate responsibility is a matter of luck; however, in moral terms, they could be urged to make additional efforts to mitigate climate change, especially if we assume that being responsible involves making sacrifices.

In addition to the division between legal liability and moral responsibility, there are other ways to conceptualise responsibility. For example, David Miller (2007, 81–109) provides an interesting alternative by distinguishing two senses of responsibility. On the one hand, *outcome responsibility* is the responsibility that people shoulder for their own actions and decisions; on the other, *remedial responsibility* acknowledges that people have responsibility to aid others in need of help. Another useful conceptualisation is Iris Marion Young's division between the liability model of responsibility and the social connection model of responsibility. 'Under this liability model', Young (2006, 116) writes, 'one assigns responsibility to a particular agent (or agents) whose actions can be shown to be causally connected to the circumstances for which responsibility is sought'. By contrast, the social connection model recognises that '[o]ur responsibility derives from belonging together with others in a system of interdependent processes of cooperation and competition through which we seek benefits and aim to realize projects' (ibid., 119).

Both Miller and Young argue that the concepts of legal and moral responsibility focus too much on causality and past actions by asking who is to blame for specific harms. In doing so, the concepts fail to identify the forward-looking responsibility of agents to seek beneficial outcomes and prevent harmful ones from happening. After all, asserting responsibility involves more than pinpointing the chief culprit in a specific crisis. As Young (2006, 122) hypothesises, the point of responsibility is 'not to blame, punish, or seek redress from those who did it [committed an act], but rather to enjoin those who participate by their actions in the process of collective action to change it'. As the principle of the responsibility to protect maintains, states have a forward-looking responsibility to *prevent* humanitarian crises whether or not they can be legally or morally held at fault for the course of events that have resulted in the current state of affairs. If states do not contribute to resolving the problems for which they are not to blame, then harmful practices will persist and could negatively affect other international practices in the process. Adhering to that system would contradict the ultimate purpose of assigning responsibilities – that is, 'not for duty-bearers to suffer more but for right-bearers to enjoy more of what they are entitled to' (Shue 1988, 697). Responsibility is therefore not only retrospective, although it largely consists of elements derived from legal and moral ethics.

Yet another conceptualisation is Robert Jackson's situational ethics, which emphasises that, in the end, taking or assigning responsibility involves making responsible choices. Responsible choices should not, as Jackson (2000, 22) notes, be 'confused with perfect choices': 'Human decisions, especially political decisions, cannot be expected to be perfect' but 'only be expected to be justified'. Accordingly, 'Responsible choices are the best choices in circumstances, or at least most defensible choices' (ibid.), meaning that when we assess state responsibility, we can at least expect states to make responsible choices in restrictive circumstances. Responsible choices are difficult decisions made between 'conflicting but equally compelling' interests and values; sometimes such choices are 'between greater and lesser evil', and sometimes they involve sacrifice (ibid., 142). Because making responsible choices is highly difficult decision making, Jackson suggests that we characterise a responsible state leader as 'somebody who can make the best of a bad situation' (ibid., 148).

Responsibilisation

As demonstrated in the previous section, responsibility is a social conception. There is no single moral compendium that applies to everybody in all circumstances or any 'final authority', comparable to God, which would have the highest moral authority. Life is far too complex to formulate a single, universal moral ethics, but statements about responsibility always derive from human practices. The only way to evaluate an actor's responsibilities is to situate them in the context of the social practice or practices in which he, she or it operates as a moral actor. By extension, the responsibilities of states are not given or known facts of life, but defining and allocating them are matters of ethics. Legal ethics makes no exception, either; viewing any legal text as a given is unnecessary because legal texts are generally reflections and products of social practices. Legal responsibility is thus always relative and cannot simply be 'read off the facts' or '*discovered*' because it is 'something to be *decided*' (Feinberg 1970, 27).

Inspired by William Clapton (2011), who has theorised processes of *riskisation*, meaning social processes in which certain issues are identified, assessed and managed as risks and their constitutive effects on international society, I apply the concept of *responsibilisation* to scrutinise processes in which (international) responsibilities are constructed. As much as issues become risks or security threats by way of riskisation or securitisation, some issues are *responsibilised* via discursive practices in order to promote their normative importance, for instance, or to oblige agents to take certain measures. My conceptualisation of responsibilisation differs significantly from the sort of responsibilisation that appears in studies of neoliberal governmentality and criminology in reference to a state's disavowal of responsibility or shunting of responsibility to its citizens.[5] By contrast, my notion of responsibilisation stresses that responsibilities are not given but socially conceived; consequently, the only way to study them is to investigate processes of how and by whom they are discursively shaped in (international) social interactions as well as how and by whom they are performed in specific settings. Understanding why someone is considered to be responsible to another for an event in a specific context requires systematically analysing numerous aspects of responsibility, especially agency, subject and object (i.e. who should be deemed responsible for what and to whom or what), normative context (i.e. which values underpin notions of responsibility), institutional organisation (i.e. whether notions of responsibility are institutionalised and, if so, then how) and the social and political effects of responsibilisation (i.e. whose notions of responsibility work to empower or marginalise others), for instance (cf. Clapton 2011).

Responsibilisation suggests that though responsibility is impossible to see or measure, it can be discussed, hence my close attention to discourses in studying processes of responsibilisation. Interviews, for example, can afford insights into what is understood to be responsible behaviour in specific contexts and why. My understanding of responsibilisation assumes that secondary institutions are key venues of the international politics of responsibility because they offer states and non-state actors a forum in which to negotiate definitions, discuss the distribution and implementation of rights and responsibilities in international society and monitor the fulfilment of those agreements. In contrast to primary institutions, secondary ones are empirically observable bodies, meaning that their decision-making procedures, rules and other organisational structures can be investigated. It is also possible to examine the power relations among members of secondary institutions and analyse the contributions of specific participants to processes of responsibilisation. Moreover, secondary institutions and their constitutive documents are principal sources for collecting empirical research material about how primary institutions have sustained and organised international society in the modern era (Kopra 2018, Kopra forthcoming).

Resolutions of the UN General Assembly provide a good starting point from which to consider the emergence and evolution of processes of responsibilisation in international society. At first glance, such abstract declarations may seem insignificant and irrelevant to actual political practices. After all, they do not necessarily create legally binding obligations but instead express what states hope to achieve. From a legal perspective, however, they are important 'acts from which views about customary law can be inferred' (Perrez 2000, 278). 'What matters', Falkner (2012, 514) observes, 'is that they represent an explicit manifestation of an implicitly assumed and broadly accepted fundamental norm'. The more often the UN reiterates the environmental responsibility of states, for instance, the more likely the UN is to affect both international law and state practices. At the same time, though changes in political discourses are integral to the process of changing political practices, focusing on discourse is not entirely sufficient for studying state responsibility. It is worthless to merely discuss responsibility because responsibility has to materialise in responsible actions both at home and abroad. When studying climate responsibility in particular, it is therefore necessary to look beyond statements expressing responsibilities, for how such words are institutionalised and acted upon requires scrutiny. Ultimately, we need to consider the broader ramifications of those processes as well.

Drivers of responsibility

Ian Hurd (1999) identifies three general reasons why states participate in international practices: coercion, calculations as well as identity and belief (cf. Buzan 2004a, 103, 130–133, 253–261; Hurd 2007, 30–40; Hurrell 2007, 67–77; Wendt 1999, 247–250). I argue that all three drivers are active processes and that, contrary to the constructivist tendency, it is unnecessary and impossible to assess the extent to which a state has internalised the rules of international practice. First, coercion, which Hurd identifies from a realist perspective, is the weakest reason of the three because social practices forcefully imposed by outsiders are not internalised whatsoever by actors themselves. Second, calculations, identified from a liberalist perspective, rests upon rational self-interest comparisons of the costs and benefits of participating in international practices. Third, identity and belief, identified from a constructivist perspective, is the most profound and stable of the three reasons;

states participate in and follow the rules of a given international practice because they believe in the moral legitimacy of the rule or the legitimacy of the international organisation that formulated it (Hurd 1999, 387). In reality, the relationship of the three drivers of internalising practices is complex, and none of them is likely to exist in a 'pure, isolated form' (ibid., 389). On the contrary, because all social practices are held together by a mélange of all three drivers, it is the 'necessity of mixture, and how to deal with it, [that] … defines politics' (Buzan 2004a, 130). Of course, states' identities and preferences change over time; the world changes, political leaders change and values change. Even if a state participates in an international practice for egoistic or other less-than-magnanimous reasons at one time, it does not mean that those interests alone will motivate the same practice in the future. At the same time, international practices can also influence participants' beliefs and identities by shaping their values and preferences. Participation in international climate practices, for example, can change a state's ideas of human wellbeing; though a state might have previously prioritised economic factors of wellbeing, it could begin to give greater value to cultivating a clean environment and a stable climate system. Over time, initially groundbreaking ideas can even become established practices taken for granted in social relations.

In the following sections I only briefly introduce Hurd's first two reasons why states participate in international practices – coercion and calculations – for the former is not a widely meaningful factor in the context of climate responsibility, whereas the latter has already been extensively studied by rationalists. By contrast, because Hurd's third reason, legitimacy and belief, is of interest to the study of state responsibility, I explore it in greater detail and from a broader perspective of identity politics. In particular, I argue the links among identity, participation in social practices and responsibility are especially strong.

Coercion and regulations

'Coercion', Hurd (2007, 35) writes, 'refers to a relation of asymmetrical physical power among agents, where this asymmetry is applied to changing the behavior of the weaker agent'. Amid the circumstance of coercion, a state participates in an international practice because it is physically or psychologically forced to. In other words, a state's participation in the practice is motivated by the fear of retribution or of physical compulsion. Coercion has been an important means of the expansion of European international society; many non-Western states, including China, were coerced into participating in international practices by way of colonisation and other oppressive means prior to the twentieth century. However, practices of solidarity, including human rights, can also be spread via coercion. For instance, in response to climate change, an eco-intervention could be made to coerce reluctant states to adopt environmentally beneficial laws.

Economic sanctions are a typical example of a non-violent form of coercion in contemporary international politics. International law can also be viewed as a more restrained and prudent form of coercion; when an international norm is given the status of law, states become legally bound to adhere to the norm. The status of law thus 'constitutes an independent reason for action' (Bodansky 2010, 91). To a similar extent, international treaties can set directives and specific responsibilities in order to guide the conduct of participants, who might be permitted, encouraged, or even required to take a certain action or not. At times, states might also be socially coerced to follow the formal rules of a practice – for example, majority decision making – because they are 'adopted in a manner that the actor accepts as legitimate' even if the states themselves resist such rules (ibid., 90). Once a state has ratified a treaty, it is obliged to follow the treaty's rules and fulfil its stipulated responsibility not because it has internalised the responsibility but because it is bound by the treaty and acting otherwise could warrant sanctions. In that case, a state's compliance is not only coerced by international regulations but also motivated by calculations. Last, international regulations additionally influence state behaviour, whether or not the state has ratified a specific international treaty concerning the issue. Even if the political leaders of a state do not place a premium on solidarist practices such as human rights and animal welfare, international practices and regulations might nevertheless constrain and influence domestic policies. In that case, the agency of non-state actors is essential to creating social pressure put upon states.

Power politics is an important incentive for states' participation in international practices. States can use international institutions to promote their values and policies globally, which is naturally considered to constitute a more legitimate means than coercion. As Franz Xaver Perrez (2000, 340) observes:

> efforts to ensure international cooperation may be conceived sometimes rather as attempts to coerce less powerful states to bring their behaviour into conformity with the interests of the most powerful states than as efforts to solve common problems cooperatively.

Although negotiations of responsibilities within secondary institutions are bound by power, the most powerful participants cannot dictate what sort of responsibilities participants ought to shoulder. All of the participants can usually participate in negotiations regarding what sort of responsibilities they are assumed to bear; in that sense, they are not coerced into taking responsibility, for such responsibilities are voluntary and not subject to coercion. It is therefore unlikely that real normative change results from coercion.

Calculations

Liberalists maintain that states cooperate with each other because it is in their interest to do so. Above all, they argue, the costs of warfare have increased, and states cannot solve global problems without international cooperation. Moreover, international norms and organisations are available to help states to solve common problems and organise cooperative efforts. From that perspective, states participate in international practices and make commitments in order to promote and maintain their national interests for the same reason that they comply with international treaties that they have signed. Because most of the goals and interests of a state are domestic, a state's participation in international practices is thus motivated by domestic interests; the needs and desires of the state guide the practices in which it participates, how it attempts to shape the goals and rules of the practice and which sorts of responsibilities it is willing to shoulder. In response to climate change, each state makes calculations regarding whether and, if so, then to what extent it is willing bear the costs of mitigating climate change and what the costs of non-participation should be in terms of lost credibility and financial consequences. Such calculations necessarily involve normative evaluations of the value that a state places upon cultivating a healthy environment. For example, does a state regard environmental protection only in terms of costs, or does nature have some intrinsic value in the calculations?

Self-interest can be an influential incentive for states to participate in international practices. Cost–benefit calculations can motivate a state to change its behaviour and assign responsibility for practices because non-participation would harm its interests and image on an international scale. However, interests do not entirely explain state behaviour. As constructivists maintain, interests are not given, and different individuals, as well as states, have different sorts of interest. '[I]nterests presuppose identities', Wendt (1999, 231) writes, 'because an actor cannot know what it wants until it knows who it is'. Ringmar (1996, 13) similarly observes, 'It is *only as some-one* that we can want *some-thing*, and it is only once we know who we *are* that we can know what we *want*.' Cost–benefit calculations are therefore unlikely to cause profound changes in an actor's preferences and values, and consequently, their influence can be brief or limited, if not both. Long-term relations among participants motivated by self-interest are thus difficult to maintain because such participants do not value those relations or sorts of cooperation. Social practices that rely heavily on self-interest necessarily have weak foundations and can easily disintegrate if power relations shift (Hurd 1999, 387). Therefore, we have to examine a state's identity in order to understand why it assumes responsibilities that can even involve undermining its interests.

Identity, practices and responsibility

Cosmopolitans believe that globalisation and increased interdependence among states and populations have fostered the emergence of a global society characterised by solidarist notions of morality. In accordance with the arguably fragile global 'we-feeling', Hurd's third reason why states participate in international practices indicates that states take part because they believe they have to. That belief relates closely to identity, which is both a subjective and an objective discourse of the self; it concerns how a person, society or state as well as others perceive and establish distinctiveness in social interactions. At the personal level, questions to that end include 'Who am I?', 'What am I?' and 'What do others think I am?' Identity is also always linked to others, the common question about which is 'What am I not?'. Furthermore, identity is material as well as ideational. Although based on the material site of a body or the territory of a state, ideas – values, beliefs, knowledge, attitudes and memories – make it special. Identity is 'what allows us to define what is important to us and what is not' (Taylor 1989, 30). 'My identity', Charles Taylor (1989, 27) explains, 'is defined by the commitments and identifications which provide the frame or horizon within which I can try to determine from case to case what is good, or valuable, or what ought to be done, or what I endorse or oppose.' Such 'commitments and identifications' are constructed within social practices – for example, religious, political, educational and family practices in which a person, society or state participates – that allocate certain responsibilities to participants. Identity is therefore a 'lived experience of participation in specific communities' (Wenger 1998, 151), and practices shape participants' identities, notions of morality and senses of appropriate courses of action. Because practices anchor identities 'in each other and what we do together', it is challenging to transform identity in isolation from other participants involved in the practice (Wenger 1998, 89). If a person is excluded from a practice important to his or her identity, he or she might face a so-called 'identity crisis'. By extension, large communities such as states need to have normative and organisational ideas that 'signify to their members what they stand for' and 'guide them in their interactions in the international arena' (Legro 2005, 6). Ideologies, for instance, are beliefs that define what is deemed right and wrong in a society (Watson 1982, 68). They are incorporated in identities and embedded in various social practices related to governmental procedures, education systems and the rhetoric of the political elite (Legro 2005, 6; cf. Haas 2005).

Again, an identity is not an exclusive, inherent asset of a person, society or state because others shape that identity as well. As Ringmar (1996, 13) explains, 'We need *recognition* for the persons we take ourselves to be, and only *as recognised* can we conclusively come to establish an identity'. Status or social identity is therefore an important element of identity, and the pursuit of a favourable one can be a significant driver for a state's participation in international practices. That idea relates to questions of self-interest and calculations discussed above; it can be quite difficult to distinguish legitimacy from interests, for states may accept international norms because doing so serves their interests. Conversely, the concept of legitimacy clarifies why states sometimes participate in practices against their own interests. According to Mark Suchman (1995, 574), *legitimacy* is 'a generalised perception or assumption that the actions of an entity are desirable, proper, or appropriate within some socially constructed system of norms, values, beliefs, and definitions'. If states accept the rules of an international practice as legitimate and justified, then they participate in the practice not only due to fear of retribution or calculations of self-interest but because of their 'internal sense of rightness and obligation' (Hurd 2007, 30). The perception of legitimacy 'may come from the substance of the rule or from the procedure or source by which it was constituted' (ibid., 381). In that case, a state internalises the rules of a practice and incorporates them into its identity and interests, and, as a result, it assumes responsibility for the practice because it is the fairest course of action to take.

Consequently, there are fundamental links among practices, identity and responsibility. The first link, between identity and responsibility in particular, is historical. To some extent, any current self-understanding is a product of past choices and commitments. For example, China's contemporary identity and approach to responsibility has been shaped strongly by its past imperial and Maoist practices. The second link concerns the here and now and, at a personal level, asks the question 'Who am I?' Usually, the answer is a name ('I'm Mary') or a statement related to practices in which a person currently takes part ('I'm the mother of Mary', 'I'm a Catholic' or 'I'm a professor'). The latter sort of answer thus refers to what Hart (1968, 212) would call *role-responsibility*, what Wendt (1999, 227) would call *role identity* and what role theorists would call *role conceptions* (e.g. Harnisch 2011). The third link is social; the commitments and identifications of others also shape self-understandings and the responsibilities that people assume. To quote Buzan (2004b, 68):

> At the end of the day, it is not what states are, or what they say about themselves and others, that determines status, but how they calculate their own behaviour and, most importantly, how they respond to the behaviour of others.

Regarding China's climate responsibility, the commitments of developed countries form an essential precondition of China's motivation to assume greater global responsibility. At the same time, there is no causal relationship between identity and responsibility, and identities alone cannot explain action. To understand why and what sort of responsibilities a person assumes in practice, we have to explore their interests. By extension, when it comes to states, national interests and goals are important when assigning responsibilities because each state – more accurately, its government – has certain goals that it seeks to accomplish. Those goals reflect a state's identity and values and motivate it to take certain actions. Accordingly, a state's identity and interests influence the sort of practices in which it might participate and what sorts of responsibilities it is willing to or capable of assuming in relation to the practice. Free-riding and the failure to fulfil those responsibilities would harm its international image and self-conception – its identity – as a responsible member of the international society.

Climate change and practices of state responsibility

Within the English School, pluralists take a highly state-centric approach to responsibility. They emphasise the values of states and pay less attention to other ethical aspects of state practices. Because pluralists regard international order as the most valuable common good of international society, they stress the duty to uphold international order as a state's primary responsibility to other states. In practice, however, a state's policies and actions, especially those of a great power, affect the lives of all people as well as non-human species worldwide. Conversely, the solidarist camp of the English School underscores human values in international relations. They maintain that a state's chief responsibility is to promote human justice at home and abroad. Since both pluralists and solidarists make important observations of and contributions to international ethics, I do not confine my theorisation of state responsibility to either one but employ Buzan's highly useful conception of state-centric solidarism. According to my state-centric solidarist reading, both pluralists and solidarists essentially agree that the ultimate referent object of state responsibility is humans; both consider peace and security to be crucial to human wellbeing. From a pluralist perspective, states should uphold international order because it is a precondition of international security and, in turn, of human wellbeing. By contrast, solidarists conceive international justice to be an important condition of human wellbeing as well. Whereas solidarists regard the entire community of humans as the referent object of state responsibility, pluralists focus on the wellbeing of citizens of individual states and scrutinise the ways in which states can fulfil their domestic responsibilities to their citizens. From the perspective of state-centric solidarism, the ultimate aim of international society is therefore to promote human wellbeing.

Although pluralism has a so-called 'ought-side', solidarism has been more purposeful in campaigning for situations that ought to be pursued in order to cultivate a fairer world. However, Buzan's (2014a, 113) conception of state-centric solidarism argues that solidarism and pluralism are not necessarily opposite poles but 'interlinked sides in an ongoing debate about the moral construction of international order'. Although Jackson can doubtlessly be categorised in the pluralist camp of the English School, he also offers exceptionally useful conceptual tools for analysing state responsibility from the perspective of state-centric solidarism. According to Jackson (2000, 170–178), governments have plural, multi-dimensional responsibilities: national ones based on realism and the promotion of national interests, international ones based on rationalism and the state's membership in international society, humanitarian ones based on revolutionism or cosmopolitanism and membership in the human race and other responsibilities to the global commons based on the idea of global trusteeship and humankind's responsibility for Earth's health.

Inspired by Buzan and Jackson, I locate practices of state responsibility on a broad spectrum of differing orientations towards moral referent objects.

Such a conceptualisation is not a matter of value judgement because I do not mean that one category is somehow more important than the others. By contrast, I intend to demonstrate that different notions of responsibility are based on different ideas about the referent objects of responsibility. At one end of the spectrum is pluralism, which focuses on states as moral referent objects. Environmental issues, apart from national environmental security, are thus largely ignored from the pluralist end, and the creation and enforcement of international norms is therefore difficult and rare. At the other end of the spectrum is ecocentrism, which gives nature moral priority, and between those two ends are state-centric solidarism and cosmopolitan solidarism. Because the two ends of the spectrum are unlikely pursuits in reality, I dismiss the pluralist end of the spectrum, at which a state is responsible only for its own survival. Instead, I argue that, at a bare minimum, states have national responsibilities and are always responsible for the wellbeing of their citizens. If not, then why would states exist at all? Instead, I discuss the other end of the spectrum – ecocentrism – which I consider to be the ultimate ought-side of the responsibility of states. An important, albeit unanswerable, question thus concerns the relationship between cosmopolitanism and ecocentrism. If viewed as the very end of the spectrum, does ecocentrism presuppose a cosmopolitan world society, or can international society be ecocentric?

State responsibilities are doubtlessly very complex. Although I distinguish four categories of general responsibilities, I do not by any means argue that the categorisation is exhaustive. A state's responsibilities can overlap and conflict; they can also shift when circumstances change. Different types of international societies have different sorts of primary institutions and practices because the 'institutions of international society are according to its nature' (Wight 1999 [1946], 111). Likewise, distinct sorts of practices have distinct ethics; even if it is accepted that environmental trusteeship has emerged as a primary institution of international society (Buzan 2004a, 186; Buzan 2014a, 161–163; Falkner 2012; Falkner & Buzan 2018; Kopra forthcoming; Palmujoki 2013), the nature of the institution depends, for instance, upon where international society falls on the spectra of pluralism–solidarism and anthropocentrism–ecocentrism. In the system of states or in a highly pluralist international society, a state is interested only in the environment within its national territory. States conceive nature as a stock of resources and thus focus on environmental concerns such as pollution, waste and the insufficiency of natural resources from a local perspective. In a more solidarist international society, by contrast, states cooperate to respond to global environmental concerns because they recognise that they cause as well as suffer from environmental harm beyond their borders. At the ultimate end of the pluralist-solidarist spectrum, states may follow ecocentric principles. Altogether, the primary institution of environmental trusteeship can be highly pluralist, yet its existence does not necessarily mean that a shared norm of climate responsibility exists. Although climate change is often viewed as a subcategory of environmental problems, it differs starkly from traditional environmental problems. Climate change is truly a global problem, and all attempts to counteract it presuppose the existence of an international society.

Similarly to Jackson, I distinguish national and humanitarian responsibilities even though they can be merged into the single category of human-centric responsibility. Because I am slightly pessimistic about the potential of states to act for reasons above and beyond pluralism, I assume that states tend to pay greater attention to the wellbeing of their citizens than to the wellbeing of other humans. The term *national responsibility* also implies that states have state-centric responsibilities, including their own survival as sovereign states. In a cosmopolitan international society, both the categories of national and international responsibility can be abandoned because nationality becomes irrelevant to assessing the nature, scope and depth of responsibility. At the same time, environmental responsibility becomes particularly difficult to demarcate. If nature is viewed as having only instrumental value for humans, then differentiating it as an independent category becomes unnecessary and only the sort of environmental aspects that pertain to a state's national, international and humanitarian responsibilities need to be discussed. In that case, the focus would fall upon the environmental security of each category. However, such a view would be too limited and would not acknowledge the intrinsic value of nature. To emphasise nature as a referent object of state responsibility, I choose to distinguish it as the ultimate end of state responsibility.

National responsibility

Although many realists have averred that a state's responsibility stops at the national border, some have begun to question that assumption by introducing ethical questions into realist research agenda (cf. Chang 2011) and maintaining that states should respect the views and interests of other nations. Although Jackson's conceptualisation of national responsibility derives significantly from classical realism it does not necessarily dismiss ethics. On the contrary, it presents an enlightened version of realism holding that a state's 'first obligation' and 'chief duty' is to preserve its national interests (Wight 1999 [1946], 95; Watson 1982, 206). For Jackson (2000, 170), national responsibility is par excellence a 'moral relation between a state and its citizens' as he sees the moral obligation of national interest as the 'fundamental standard of conduct' and national security as a foundational value. In other words, Jackson suggests that national interest is a 'moral idea governing the conduct of statespeople: the idea that the nation and its population are a treasure which they have the responsibility to safeguard' (ibid., 21). Therefore, states have a moral obligation to defend national interests driven by a 'moral concern for the flourishing of the national population, for their good life' (ibid.,171). Clearly, however, national interest can be viewed as a moral guideline only if defined more broadly than in narrow, self-interested terms that focus on state security.[6]

From another angle, literature on happiness suggests that governments face strong incentives to assume the happiness of citizens as their ultimate responsibility (e.g. Duncan 2010; Bok 2010). However, thus far, Bhutan is the only nation to have adopted 'gross-national happiness' as the central aim of its national policy.[7] A pluralist approach to national responsibility would suggest that the happiness of the state – or more broadly, the wellbeing of citizens – should be the only legitimate goal of national policies. However, solidarists would disagree for both moral and practical reasons. They would argue, for example, that even if a government's moral duty is to promote the wellbeing of its citizens, it should not do so by infringing upon the wellbeing of citizens of other states because it has responsibilities beyond its own borders. Moreover, they would add, responsible governments should not exclusively promote the short-term wellbeing of their present populations and not avoid making difficult decisions that promote the long-term wellbeing of their citizens but conflict with their short-term (e.g. economic) interests.[8]

Because the concept of national responsibility emphasises the significance of the domestic responsibilities of states, it identifies international law and secondary institutions as 'instrumental arrangements which are justified by how well they serve the national interest of states' (Jackson 2000, 170). Accordingly, the concept maintains that states are foremost responsible for the wellbeing of their own citizens, not that of foreign countries and their populations (ibid., 171). As for foreign affairs, that sort of normative standard supports Machiavellian principles of self-interest because it holds that a state needs to put its own nation and citizens first and cooperate with other states only when necessary to promote national interests and, among other things, to avoid putting their citizens and military at risk of harm. Linklater and Suganami (2006, 235) rightly criticise Jackson's (2000) argument for its incomparable normative basis for national responsibility. According to Jackson, the '"first duty of a government is to protect its own people. After that it can try to help whoever else it can"' (quoted in Linklater & Suganami 2006, 235). Taken literally, Jackson's argument means that states have to first do whatever they can to assist their own citizens and only afterwards may they pay attention to the needs and interests of citizens of other states. However, Linklater and Suganami admit that such an idea is not necessarily the intention of Jackson's argument, for it would mean that, during a war, a state could ignore the international codification of humanitarian law and do whatever it pleased to secure the interests of its people and only later, if at all, think about the human suffering of the soldiers and civilians of opposing states (ibid.). Because Machiavellian principles were defined amid a system of separate and often rival states, it is understandable that they consider state responsibilities in purely national terms. International society did not exist at the time, and there were no responsibilities beyond a state's national borders; if there were, then they derived from the national interest of securing peaceful international order. In today's global era, however, such principles are inevitably outdated and do not provide a normative basis for international ethics.

Sovereign states, at least liberal democratic ones, define and allocate their national responsibilities according to their constitutions and other laws. However, such practices do not mean that states are responsible for everything that occurs within their borders; after all, states are not protectors of property and life, nor are they responsible for their citizens' actions. Entering into international agreements does not decrease a state's sovereignty; it may even preserve that sovereignty when international cooperation is needed to diffuse transnational threats that risk its sovereignty. The concept of national responsibility does not necessarily ignore the environment, either. Unlike traditional approaches to security that focus on a state's national security, a broader approach is concerned with human and environmental security and emphasises the idea that states bear a responsibility to protect their citizens from threats of environmental origin. It is becoming increasingly more certain that causal links therefore exist between environmental degradation and inter- and intra-state conflicts.

International responsibility

Based on liberalism, Jackson's conceptualisation of international responsibility suggests that, unlike the Hobbesian 'war of all against all', states form an international society. Such an international society is constitutional in nature, for its members' duties and rights are written into international law, among which the UN Charter (1945) is the most fundamental. When international society recognises a state's sovereignty and membership in such a society, the state presumably accepts and becomes capable of exercising its rights and responsibilities in that society. It is, to quote Eagleton (1928, 5), 'upon this agreement to observe the rules of the community that international responsibility is founded'. Given the constitutional relationship, states are not only accountable to their own citizens but also 'responsible for upholding international law and the society of states as a whole' (Jackson 2000, 172). They have a general responsibility to preserve international society and maintain its functioning, and they are obliged to pursue their national responsibilities without infringing upon the rights of other states. Due to the state-centric basis of international society, the most essential international right listed in the UN Charter is non-intervention, meaning a responsibility to not intervene unless in lawfully recognised circumstances. States also have a right and responsibility to participate in diplomatic practices, meaning that, at a minimum, every state should respect the UN Charter, prevent harm to others and refrain from unnecessary military action. However, if a state fails to uphold its responsibilities, methods of sanctioning states in the absence of a supranational body remain unavailable.

From the perspective of the English School, international law provides an important normative framework no less binding than domestic law within which and with reference to which states make choices about what actions to perform in international society. Apart from the Responsibility of States for

Internationally Wrongful Acts drafted and compiled by the International Law Commission in 2001, however, there is no international law regarding state responsibility. From the perspective of international law, states can be held responsible for pollution only if such pollution is wrongful under international law. Because carbon dioxide and other greenhouse gas emissions are legal forms of pollution caused primarily by the activities of individuals and private corporations, states cannot be held legally responsible for the damage caused. Conversely, the idea of a state's international responsibility usually refers to its political responsibilities as the most authoritative within international society.

As members of international society, all states have equal, general responsibilities derived from the UN Charter. They have a responsibility to safeguard international peace and security, prosperity and the wellbeing of people of present and future generations both locally and globally. They are also obliged to not cause harm to others. In practical terms, however, the circumstances and capacities of states vary considerably, as does their power and capability to shoulder international responsibilities. Henry Shue's (1993) dichotomy between the general responsibilities of all states and the special responsibilities of states with greater capabilities elucidates the practices of global responsibility, including those related to the climate. Regarding climate responsibility, Shue's distinction between *subsistence emissions* and *luxury emissions* clarifies that not all greenhouse gas emissions are equally detrimental. In short, developing countries' subsistence emissions, or *survival emissions*, are unavoidable because they are produced in order to guarantee a basic minimum standard of living for the poor. Whereas general responsibilities relate to so-called 'arithmetical justice', meaning that states have equal rights and responsibilities, the distribution of special responsibilities is a political decision made by international society as a whole in the 'consideration of its common good or interest' (Bull 2002 [1977], 77). Although states generally agree that the distribution of responsibilities is a matter of fairness and some states have special responsibilities, political debate about the ethical underpinnings of ways to define and distribute responsibilities equitably remains heated. From another perspective, moral philosophers have elaborated upon the fairness of the allocation of climate responsibilities (e.g. Caney 2010; Gardiner 2011; Gardiner, Caney, Jamieson & Shue 2010; Vanderheiden 2008).

From a legal and moral perspective, a significant causal link joins contribution and responsibility. If a person, society or state is guilty of an action, then it has a responsibility to remedy that action's effects. In international environmental politics, that notion has been best captured by the so-called 'polluter pays' principle, which is nevertheless problematic at least for three reasons. Many polluters cannot pay because they are dead, many simply cannot afford to pay and many refuse to pay (Caney 2010, 134). Although it is impossible to pinpoint who is guilty of causing climate change, which is caused by countless people participating in morally and legally accepted practices, they can be collectively held responsible for it.[9] However, would it be fair to ask all states or people to shoulder similar responsibility to mitigate climate change?

Humanitarian responsibility

Jackson's (2000) idea of humanitarian responsibility derives from cosmopolitanism and the notion of world society. Cosmopolitanism maintains that people have universal negative responsibilities not to dispossess other people's rights (Shue 1988, 690). Due to their specific positions and capacity to improve or harm the wellbeing of fellow humans around the world, state leaders are responsible for the good life of all people, not only the citizens of their respective countries (Jackson 2000, 174–175). Therefore, states have a fundamental positive responsibility to 'respect the dignity and freedom of human beings' and need to do their utmost to defend human rights all over the world. From a humanitarian standpoint, 'respect for human beings – whoever they are and whatever they happen to be living – is a fundamental normative consideration in foreign policy' (ibid., 174). At the same time, the humanitarian approach is also deeply rooted in Western traditions, including Christianity. From the Chinese perspective, conversely, it is entirely Eurocentric to justify the humanitarian responsibilities of states by invoking natural law and the universal valuation of human rights.

International practices emphasising human rights began to evolve after the atrocities committed during World War II, when liberal-democratic states agreed upon new humanitarian principles for international society. Procedurally, the humanitarian approach is now written into international humanitarian law. The Universal Declaration of Human Rights (1948) recognises that all humans are born free and have an equal right to life, liberty and security. In practical terms, the humanitarian approach is recognisable in the strong human rights doctrines of Western countries. Linklater and Suganami (2006, 243) regard Article 5 of the Universal Declaration of Human Rights, which states that all humans have the right to be free from 'torture, cruel, inhuman or degrading treatment or punishment', as the '*grundnorm* of the solidarist position on good international citizenship'. During the Cold War, questions of human right norms were muted in practice. In more recent decades, however, interest in international humanitarian ethics has rapidly risen as globalisation has introduced new sorts of political concerns, including global inequality and justice, sustainable development and non-state actors' increased role in world politics. Although those global developments indicate that states 'cannot advance significantly beyond a pluralist conception of good citizenship', there is nevertheless room for solidarist ethics (ibid., 229).

In particular, the concept of sovereignty as responsibility developed by Francis M. Deng et al. (1996) has stimulated extensive political and academic discussions about state responsibility that have prompted a shift from *the right to interfere* to the *responsibility to protect* (e.g. Wheeler 2000). In 2001, the International Commission on Intervention and State Sovereignty (2001, XI) introduced the 'responsibility to protect', for which it highlighted two basic principles: that 'state sovereignty implies responsibility, and the primary responsibility for the protection of its people lies with the state itself' and that 'where a population is suffering serious harm, as a result of internal war, insurgency, repression or state failure, and the state in question is unwilling or unable to halt or avert it, the principle of non-intervention yields to the international responsibility to protect'. The UN General Assembly (2005) adopted the 'responsibility to protect' as follows:

> Each individual State has the responsibility to protect its populations from genocide, war crimes, ethnic cleansing and crimes against humanity. This responsibility entails the prevention of such crimes, including their incitement, through appropriate and necessary means. We accept that responsibility and will act in accordance with it. The international community should, as appropriate, encourage and help States to exercise this responsibility and support the United Nations in establishing an early warning capability.

The UN General Assembly (2005) has defined the *responsibility to protect* to consist of three pillars: a state's responsibilities to its citizens, international society's responsibility to assist states to fulfil their responsibilities and international society's responsibility to take collective action if a state manifestly fails to protect its citizens. That definition reveals two important aspects of state responsibility. On the one hand, sovereignty remains the fundamental principle of international society. Indeed, the definition confirms that sovereignty is an essential precondition of state responsibility, for without independence from external control and full authority over a territory, a state cannot exercise full responsibility. Therefore, the 'exclusive territorial jurisdiction of the state', as Eagleton (1928, 7) writes, 'is the chief source of its responsibility'. On the other, the definition indicates that state responsibility presupposes the existence of international society and is therefore constructed in social interaction. After all, if there were only one state, then the concepts of external sovereignty and the responsibility to protect would not make much sense.

Climate change violates basic human rights, including the right to life, the right to health and the right to subsistence (Caney 2009, 230–231). The Human Rights Council of the UN acknowledged that dilemma in 2008 when it passed its first resolution related to climate change, according to which 'climate change poses an immediate and far-reaching threat to people and communities around the world and has implications for the full enjoyment of human right' (Human Rights Council 2008). Because climate change endangers the traditional practices of indigenous people and the very existence of island nations, legal cases against developed countries for violating indigenous communities' human rights by causing climate change have already begun to emerge.[10] An important breakthrough in state climate responsibility occurred in 2015, when a court in the Hague ordered the Dutch government to cut emissions at least by 25 per cent within five years in order to protect its citizens from climate change.

Since carbon dioxide, the chief human cause of climate change, is a so-called 'stock pollutant', meaning that today's emissions might not harm us today but could cause problems for future generations, humanitarian responsibility can be extended to include future generations as well. Although the idea that people are concerned with the lives of future generations is nothing new, the capacity of the present generation to negatively affect the wellbeing of future generations is relatively novel. Although we cannot know with any certainty what the interests of future generations will be, we can assume that some basic needs, including access to clean water and air, are common to all humans and other animals regardless of time and place. Furthermore, if we agree that future generations have corresponding rights, then the present generation has responsibilities to them. After all, if contemporary practices harm the basic interests of future generations, then they violate their rights, and no optimism about the future's advanced technologies to clean today's pollution can reduce the responsibility of the present generation. The principle of sustainable development clearly acknowledges that fact, at least in principle. In reality, however, intergenerational responsibilities are often not discussed in terms of humanitarian responsibility (cf. Weiss 1989).

Environmental responsibility

According to Jackson (2000, 175–178), people have a conservationist responsibility because the health of Earth is vital to humans. Because people can live without nation-states but not without the planet, we are bound to shoulder responsibility for the global environment. That norm has been best captured by the idea of global trusteeship, which holds that because humans have the industrial power to shape the balance of nature, they also have a responsibility to conserve it. The greatest 'responsibility for the global commons' falls to governments, which are the 'chief trustees or stewards of the planet' because they have juridical power to regulate activities and control potential harm to the environment (ibid). From Jackson's pluralist perspective, states are expected to protect nature within their jurisdictions and take international action to preserve the global environment.

Although I do not disagree with Jackson's idea of responsibility for the global commons in general, it needs to be revisited for two reasons. First, the definition of *global commons* is not clear enough. Often, the *global commons* refers to the oceans, the atmosphere, the ozone layer, global biodiversity, outer space, the North Pole and Antarctica, of which global biodiversity is the least clear, particularly in respect to its location. Apart from ecosystems in the oceans, Antarctica and the North Pole, biodiversity is not global in a physical sense but always exists in the territory of specific countries. For example, the Amazon rainforest is typically described as a global commons despite its location in the territory of several sovereign states. People who happen to live in those countries have the privilege to decide what to do with the rainforest, how to use its biodiversity as natural resources and how to treat specific species. Consequently, as a local resource, biodiversity is governed according to a sovereign state's political, cultural and social values, principles and norms. Second, it seems that Jackson's conceptualisation of global trusteeship is highly anthropocentric. Because it suggests that humans, particularly state leaders, are responsible for the health of the planet as the only home we ultimately have, humans are the sole objects of moral responsibilities, whereas nature has no intrinsic value (i.e. the value of ends, or nature for its own sake) but only instrumental value (i.e. the value of means, or nature in terms of resources). From that perspective, states have to shoulder responsibility for the global commons only because it is in the interest of humans. By contrast, considering the intrinsic value of nature would indicate that nature must be respected and preserved for its own sake and that states have responsibilities to the natural world as such. Nevertheless, the instrumental value judgement of nature does not necessarily mean that humans have no responsibility to nature but face a strong instrumental incentive to protect nature because it provides ecosystems vital to humans.

Ecocentrism recognises that humans and therefore states have a moral standing vis-à-vis Earth and not only for anthropocentric and instrumental reasons. Ecocentrism recognises various human interests related to the environment and thus national interests as well. It does not overlook the role of anthropocentric interests such as economic needs for natural resources and human welfare, but it does provide a more encompassing approach by also recognising the interests of non-human species, ecological communities and future generations of both humans and non-humans. At a fundamental level, the conservation of biodiversity is a moral recognition that all species have rights and should not be treated as lifeless or valueless objects (Eckersley 1992).

The ecocentric conceptualisation of state responsibility is not merely a utopian idea. Practices of environmental constitutionalism, which regard the environment as a 'proper subject for protection in constitutional texts and for vindication by constitutional courts', are now evolving worldwide (May & Daly 2015, 1–2), and a growing number of states explicitly recognise the substantive environmental rights of citizens and the government's environmental responsibilities (ibid., Appendices A and C). Although such environmental rights and duties are largely motivated by anthropocentric reasons, a more ecocentric constitutionalism 'advancing the right of nature' is not uncommon (ibid., 255).[11] Plus, contemporary international law 'already carries the seeds of possibility for non-anthropocentric conceptions' of responsibility (Bosselmann 2015, 40), as I demonstrate in the following chapters. Currently, however, international environmental law is highly fragmented and largely ill equipped to fulfil climate responsibility (cf. Voigt 2008), and no independent international treaty on environmental rights exists. In fact, the UN International Law Commission's State Responsibility Project in the 1980s and 1990s abandoned the idea that state criminal responsibility includes the widespread pollution of the seas and atmosphere (Koivurova 2014, 174–175). At present, the most prominent articulation of ecocentric responsibility appears in the Earth Charter (2000), a civil society initiative launched in 2000, whose first principle urges humans to 'Respect Earth and life in all its diversity', based on an ontological assumption that 'all beings are interdependent and every form of life has value regardless of its worth to human beings'. However, because the Earth Charter is not endorsed by states, its legal status remains a document of soft law and is not legally binding for governments. Nevertheless, because it is a statement expressed by the world over civil society, it is, to use Klaus Bosselmann's (2015, 110–111) words, 'arguably one of the most legitimate international statements of principle ever to be made'. As such, it should be morally binding to states.

Conclusion

Both the legal and moral conceptualisations of responsibility are pivotal for analysing the scope of state responsibility, especially from a historical perspective. However, they do not pay attention to the fact that international responsibilities are not static but produced and reproduced in social interaction. Because they do not recognise that states can fulfil their forward-looking responsibilities only by joining with others in the practices of international society, legal and moral approaches to responsibility are insufficient views for studying state responsibility (cf. Young 2006, 123). This chapter has demonstrated that responsibility not only looks retrospectively at the past even if it largely consists of elements derived from legal and moral ethics. In that regard, the English School's theorisation of responsibility has proven to be valuable. This chapter has also demonstrated that state-centric solidarism offers an enlightened approach for conceiving state responsibility by acknowledging that, in addition to inter-state responsibilities, governments are always responsible for the wellbeing of individuals. From a thinner, pluralist perspective, they are chiefly responsible for the citizens of their own states, whereas from a thicker, solidarist perspective, they have responsibilities for the wellbeing of all humans worldwide. The thickest perspective acknowledges that states also have responsibilities for the wellbeing of planet Earth.

Notes

1 See Shue (1980) for a detailed study of basic rights and Alfredsson (2010), Anton and Shelton (2011) and Weiss (1989, 95–117), for example, for comprehensive analyses with numerous examples of how the environment has been treated in the field of human rights and how international environmental regimes incorporate human rights.
2 For a detailed analysis of legal responsibility in the context of civil and criminal law, see Fletcher (1998), Ross (1975), Hart (1968) and Morris (1961).
3 For a detailed elaboration on causation and responsibility, see Hart and Honoré (1985).

4 By contrast, China's climate policies are sometimes criticised because they are not implemented for the so-called 'right' reasons. For example, Richerzhagen and Scholz (2008, 311) complain that though renewable energy has been priortised in China's energy strategy since 2006, related measures may not have been implemented for the climate's sake but to cut energy costs and increase energy security, which are essential measures to maintain economic growth.

5 According to Alison Wakefield and Jenny Fleming in the *Sage Dictionary of Policing, responsibilization* is a 'term developed in the governmentality literature to refer to the process whereby subjects are rendered individually responsible for a task which previously would have been the duty of another – usually a state agency – or would not have been recognized as a responsibility at all. The process is strongly associated with neoliberal political discourses, where it takes on the implication that the subject being responsibilized has avoided this duty or the responsibility has been taken away from them in the welfare state era and managed by an expert or government agency'.

6 *Public opinion* is not synonymous with *national responsibility*; at times, the fulfilment of the latter requires difficult decisions and policies that conflict with the former. In democratic societies, politicians might fail to shoulder their national responsibilities because they have to consider voters' opinions and preferences in order make a case for their re-election.

7 According to Bok (2010), Bhutan's four pillars of gross national happiness are good governance and democratization, stable and equitable socioeconomic development, environmental protection and the preservation of culture.

8 Environmental protection is an excellent example, for it sometimes conflicts with people's short-term interests, especially those measured in economic terms. Similarly, banning cigarettes would promote people's health and thus their happiness. In the short term, however, people would probably not be pleased with the ban, and their happiness might even diminish.

9 Larry May's (1992, 38) distinction between shared and collective responsibility holds that '[w]hen a group of people shares responsibility for a harm, responsibility distributes to each member of the group. When a group is collectively responsible for a harm, the group as such is responsible; but this does not necessarily mean that all, or even any, of the members are individually responsible for the harm.'

10 See, for example, the *Petition to the Inter American Commission on Human Rights Seeking Relief from Violations Resulting from Global Warming Caused by Acts and Omissions of the United States* (2005) and the *Petition to the Inter American Commission on Human Rights Seeking Relief from Violations of the Rights of Arctic Athabaskan Peoples Resulting from Rapid Arctic Warming and Melting Caused by the Emission of Black Carbon by Canada* (2013).

11 For example, Ecuador's constitution includes a chapter on nature's rights.

Bibliography

Alfredsson, Gudmundur. 2010. 'Human rights and the environment'. In David Leary & Balakrishna Pisupati (eds), *The Future of International Environmental Law*. Tokyo, New York, Paris: United Nations Press, 127–146.

Anton, Donald K. & Dinah L. Shelton. 2011. *Environmental Protection and Human Rights*. Cambridge: Cambridge University Press.

Bodansky, Daniel. 2010. *The Art and Craft of International Environmental Law*. Cambridge, MA: Harvard University Press.

Bok, Derek. 2010. *The Politics of Happiness: What Government Can Learn from the New Research on Well-being*. Princeton and Oxford: Princeton University Press.

Bosselmann, Klaus. 2015. *Earth Governance: Trusteeship of the Global Commons.* Cheltenham: Edward Elgar Publishing.

Bull, Hedley. 2002 [1977]. *The Anarchical Society: A Study of Order in World Politics*, 3rd edition. Basingstoke: MacMillan Press.

Buzan, Barry. 2014. *An Introduction to the English School of International Relations.* Cambridge: Polity Press.

Buzan, Barry. 2004a. *From International to World Society? English School Theory and the Social Structure of Globalisation.* Cambridge: Cambridge University Press.

Buzan, Barry. 2004b. *The United States and the Great Powers: World Politics in the Twenty-first Century.* Cambridge: Polity Press.

Caney, Simon. 2010. 'Cosmopolitan Justice, Responsibility, and Global Climate Change'. In Stephen M. Gardiner, Simon Caney, Dale Jamieson & Henry Shue (eds), *Climate Ethics: Essential Readings.* Oxford: Oxford University Press, 122–145.

Caney, Simon. 2009. 'Human rights, responsibilities, and climate change'. In Charles R. Beitz & Robert E. Goodin (eds), *Global Basic Rights.* Oxford: Oxford University Press.

Chang, Chih-Hann. 2011. *Ethical Foreign Policy? US Humanitarian Interventions.* Farnham: Ashgate.

Clapton, William. 2011. 'Risk in international relations'. *International Relations* 25:3, 280–295.

Deng, Francis M. et al. 1996. *Sovereignty as Responsibility: Conflict Management in Africa.* Washington: Brookings Institute.

Duncan, Grant. 2010. 'Should happiness-maximization be the goal of government?'. *Journal of Happiness Studies* 11:2, 163–178.

Eagleton, Clyde. 1928. *The Responsibility of States in International Law.* New York: New York University Press.

Earth Charter. 2000. Accessed 3 March 2018. http://earthcharter.org/.

Eckersley, Robyn, 1992. *Environmentalism and Political Theory: Toward an Ecocentric Approach.* London: UCL Press.

Erskine, Toni. 2003. 'Assigning responsibilities to institutional moral agents: The case of states and 'quasi-states'. In Toni Erskine (ed.), *Can Institutions Have Responsibilities?* Basingstoke: Palgrave Macmillan, 19–40.

Falkner, Robert & Barry Buzan. 2018. 'The emergence of environmental stewardship as a primary institution of global international society'. *European Journal of International Relations.*

Falkner, Robert. 2012. 'Global environmentalism and the greening of international society'. *International Affairs* 88:3, 503–522.

Feinberg, Joel. 1970. *Doing and Deserving.* Princeton: Princeton University Press.

Fletcher, George P. 1998. *Basic Concepts of Criminal Law.* New York: Oxford University Press.

French, Peter A. 1984. *Collective and Corporate Responsibility.* New York: Columbia University Press.

French, Peter A. & Howard K. Wettstein (eds). 2006. *Shared Intentions and Collective Responsibility.* Midwest Studies in Philosophy vol. XXX. Boston, MA: Blackwell Publishing.

Gardiner, Stephen M. 2011. *A Perfect Moral Storm: The Ethical Tragedy of Climate Change.* Oxford: Oxford University Press.

Gardiner, Stephen M., Simon Caney, Dale Jamieson & Henry Shue (eds) 2010. *Climate Ethics: Essential Readings.* Oxford: Oxford University Press.

Goodin, Robert E. 1995. *Utilitarianism as a Public Philosophy.* New York: Cambridge University Press.

Haas, Mark L. 2005. *The Ideological Origins of Great Power Politics, 1789–1989.* Ithica, NY: Cornell University Press.

Harnisch, Sebastian. 2011. 'Role theory: Operationalization of key concepts'. In Sebastian Harnisch, Cornelia Frank & Hans. W. Maull (eds), *Role Theory in International Relations: Contemporary Approaches and Analyses.* New York: Routledge, 7–15.

Hart, H. L. A. 1968. *Punishment and Responsibility: Essays in the Philosophy of Law.* Oxford: Clarendon Press.

Hart, H. L. A. & Tony Honoré. 1985. *Causation in the Law,* 2nd edition. Oxford: The Clarendon Press.

Human Rights Council. 2008. 'Human rights and climate change'. Resolution 7/23, 28 March.

Hurd, Ian. 2007. *After Anarchy: Legitimacy and Power in the United Nations Security Council.* Princeton: Princeton University Press.

Hurd, Ian. 1999. 'Legitimacy and authority in international politics'. *International Organization* 53:2, 379–408.

Hurrell, Andrew. 2007. *On Global Order: Power, Values, and the Constitution of International Society.* Oxford: Oxford University Press.

International Commission on Intervention and State Sovereignty. 2001. 'The Responsibility to Protect'. Accessed 23 March 2017. responsibilitytoprotect.org/ICISS%20Report.pdf.

Jackson, Robert H. 2000. *The Global Covenant.* Oxford: Oxford University Press.

Koivurova, Timo. 2014. *Introduction to International Environmental Law.* New York: Routledge.

Kopra, Sanna. Forthcoming. 'China and the UN climate regime: Climate responsibility from an English School perspective'. *Journal of International Organizations Studies.*

Kopra, Sanna. 2018. 'China, Great Power Management, and Climate Change: Negotiating Great Power Climate Responsibility in the UN'. In Tonny Brems Knudsen & Cornelia Navari (eds), *International Organization in the Anarchical Society: The Institutional Structure of World Order.* New York: Palgrave Macmillan.

Legro, Jeffrey W. 2005. *Rethinking the World: Great Power Strategies and International Order.* Ithaca, NY: Cornell University Press.

Linklater, Andrew & Hidemi Suganami. 2006. *The English School of International Relations: A Contemporary Reassessment.* New York: Cambridge University Press.

May, James & Erin Daly. 2015. *Global Environmental Constitutionalism.* Cambridge: Cambridge University Press.

May, Larry. 1992. *Sharing Responsibility.* Chicago: University of Chicago Press.

Mayer, Hartmunt & Henri Vogt (eds). 2006. *A Responsible Europe? Ethical Foundations of EU External Affairs.* Basingstoke: Palgrave Macmillan.

Miller, David. 2007. *National Responsibility and Global Justice.* Oxford: Oxford University Press.

Morris, Herbert. 1961. *Freedom and Responsibility.* Stanford: Stanford University Press.

Nagel, Thomas. 1979. *Mortal Questions.* Cambridge: Cambridge University Press.

Palmujoki, Eero. 2013. 'Fragmentation and diversification of climate change governance in international society'. *International Relations* 27:2, 180–201.

Perrez, Franz Xaver. 2000. *Cooperative Sovereignty: From Independence to Interdependence in the Structure of International Environmental Law.* The Hague: Kluwer Law International.

Richerzhagen, Carmen & Imme Scholz. 2008. 'China's capacities for mitigating climate change'. *World Development* 36:2, 308–324.

Ringmar, Erik. 1996. *Identity, Interest and Action: A Cultural Explanation of Sweden's Intervention in the Thirty Years War.* Cambridge: Cambridge University Press.

Ross, Alf. 1975. *On Guilt, Responsibility and Punishment.* London: Stevens & Sons.

Shue, Henry. 1993. 'Subsistence emissions and luxury emissions.' *Law & Policy* 15:1, 39–59.

Shue, Henry. 1988. 'Mediating duties'. *Ethics* 98:4, 687–704.

Shue, Henry. 1980. *Basic Rights.* Princeton: Princeton University Press.

Suchman, Mark C. 1995. 'Managing legitimacy: Strategic and institutional approaches'. *Academy of Management Review* 20:3, 571–610.

Taylor, Charles. 1989. *Sources of the Self: The Making of the Modern Identity.* Cambridge: Cambridge University Press.

UN General Assembly. 2005. '2005 World Summit outcome'. A/RES/60/1, 16 September.

Vanderheiden, Steve. 2008. *Atmospheric Justice: A Political Theory of Climate Change.* Oxford: Oxford University Press.

Voigt, Christina. 2008. 'State responsibility for climate change damages'. *Nordic Journal of International Law* 77, 1–22.

Wakefield, Alison & Jenny Fleming. 'Responsibilization.' *The SAGE Dictionary of Policing.* Accessed 3 March 2018. http://dx.doi.org/10.4135/9781446269053.n111.

Watson, Adam. 1982. *Diplomacy: The Dialogue Between States.* London: Eyre Methuen.

Weiss, Edith Brown. 1989. *In Fairness to Future Generations: International Law, Common Patrimony, and Intergenerational Equity.* Tokyo: United Nations University.

Wendt, Alexander. 1999. *Social Theory of International Politics.* Cambridge: Cambridge University Press.

Wenger, Étienne. 1998. *Communities of Practice: Learning, Meaning and Identity.* Cambridge: Cambridge University Press.

Wheeler, Nicholas J. 2000. *Saving Strangers: Humanitarian Intervention in International Society.* Oxford: Oxford University Press.

Wight, Martin. 1999 [1946]. *Power Politics.* Edited by Hedley Bull and Carsten Holbraad. London: Leicester University Press.

Young, Iris Marion. 2006. 'Responsibility and global justice: A social connection model'. In Paul, EllenFrankel, Fred D.Miller, Jr., & Jeffrey Paul (eds), *Justice and Global Politics.* Cambridge: Cambridge University Press, 102–130.

3 Practices of state responsibility in China

We cannot understand the present without understanding history. Therefore, in this chapter I briefly introduce China's political history from the perspective of state responsibility. Although I examine China's identity, values, interests and policies within a historical continuum, I do not pretend to give a comprehensive or even overall view of China's long history. On the contrary, I seek to succinctly review how China's central government has constructed notions of responsibility over time and to analyse how those notions have guided political practices in China. I base my understanding on a selective reading of practices of Chinese political leadership and developments that might influence current notions of climate responsibility in China. A historical background can elucidate what sort of responsibilities China's government is willing to or can assign to international climate politics today.

Legacy of China's imperial and Maoist eras

In ancient times, the sovereign territory now known as China was not called 'China' by people living there but *Tianxia* (天下), meaning 'all under heaven'. The Chinese emperor, regarded as the centre of the universe and the son of heaven, was considered to rule everything that mattered and was responsible for maintaining unity and stability in the empire. As the most powerful state and most advanced civilisation of the period, China was the political, economic and cultural model for other societies in the region. It formed the Chinese world order – the Pax Sinica – and defined norms of international order regionally and well before the birth of European international society (Zhang 2014, 56). Although European state and non-state actors participated in the practices of the Pax Sinica in the sixteenth century, they did not attempt to impose European practices upon China (ibid., 72). Indeed, they did not 'challenge, if ever they did question, the assumptions, worldviews, the legitimacy and predominance of the Chinese world order they encountered' (ibid., 73). The era of the Pax Sinica ended with the beginning of so-called 'Century of Shame' or 'Century of Humiliation' (1839–1949), which encompassed several lopsided treaties and losses of sovereignty, including the cession of Hong Kong to Britain. At that time, China's international responsibility was 'largely to make sacrifice as a colony', and the failed state could not fulfil the 'fundamental responsibility of maintaining its people's living standards, safeguarding basic human rights and developing economy' (Jin 2011, 7). The Century of Shame concluded with the communist revolution and establishment of the People's Republic of China (PRC) in 1949.

Before the globalisation of Western ideas of human–nature relationships in the nineteenth and twentieth centuries, the Chinese language did not include any word equivalent to *nature* (Weller 2006, 13). In ancient China, nature itself did not exist, although human–nature relationships were 'reflected and disputed' in the terms *tian* (天), *xing* (性), *sheng* (生) and others (Schmidt-Glintzer 2010, 526), and only in the early twentieth century did *ziran* (自然) become the accepted translation for the Western concept of nature (Weller 2006, 13). In contrast to the Western belief that God created nature – that is, *natura naturata* – the ancient Chinese conceived nature as the 'ever-productive and self-renewing forces of nature', or *natura naturans* (Kubin 2010, 517). As agricultural economies, both the Ming (1368–1644) and Qing (1644–1911) dynasties followed the principles of Confucianism and highly valued a balanced relationship between humans and land or humans and nature (天人关系, 人地关系, 人大自然关系, 天人合一). When rapid environmental change commenced in China during the late seventeenth century as the country's population burgeoned, the growing population increasingly used natural resources (Sit 2010, 241–242). By 1800, China faced a 'pre-modern energy crisis' caused largely by heavy deforestation (Marks 2012, 6), and the modernisation of agriculture and industry, population growth and commercialisation during the late Qing dynasty and the Republic of China (1912–1949) caused serious environmental degradation and social problems, including deforestation, energy shortages, droughts, famines and conflicts over water and land control (ibid., ch. 6). In particular, the mid-nineteenth century marked the 'watershed between traditional China and the modern age, between the pre-industrial and post-industrial age, and between the choice of neo-Confucianism and a reformed way of managing man–land relationship with regard to new technology and other factors in the modern era' (Sit 2010, 241–242). Due to social and environmental changes during the mid-nineteenth century, the PRC 'inherited a seriously degraded natural environment' when it emerged in 1949 (Marks 2012, 263).

Although the establishment of the PRC marked the beginning of a new era in China's history, events in China's past instilled in Mao Zedong, the first chair of the PRC, an obsession with sovereignty, a fear of invention and a suspicion of foreigners. Maoist China (1949–1978) resisted both formal and informal international organisations, which it regarded as 'the creations of either the superpowers or the Western capitalist camp' (Deng 2008, 4). According to Zhao Suizheng (2000, 4), all Chinese leaders in the twentieth century shared 'a deep bitterness at China's humiliation' and resolved to 'restore China to its rightful place as a great power'. Inspired by Karl Marx, Mao applauded technological progress, economic growth and the conquest of nature.[1] He collectivised most of China's land and forests in an effort to maximise industrial output without paying attention to environmental impacts. Factories had no incentives to protect the environment because they were evaluated in terms of industrial output and economic growth. Due to inefficient production technologies and heavily subsidised energy prices, energy efficiency in China was and remains poor, and natural resources have been systematically wasted (cf. Shapiro 2001). Because the volume of workers was considered to be vital to maximising industrial input, population policy was an integral part of Maoist China's national economic development, and rapid population growth was therefore heavily promoted.

In rejecting longstanding ideas of human–nature relationships based on traditional neo-Confucianism, Maoist society pursued social modernisation focused on heavy industry, forced industrialisation and central planning. In contrast to the Western view of nature as 'common heritage' or 'private property that needed to be preserved', the environment in China was regarded as a 'common good that could be put to positive use' (Kobayashi 2005, 90). Indeed, Maoist China believed that the 'utilization of natural resources is in principle inexhaustible, or at least, constantly extending' (Greenfield 1979, 217). With the militarisation of other sectors of life and the state's fight against capitalism, individualism, imperialism, feudalism and revisionism, Mao declared a 'war against nature' and promoted the slogan 'Man must conquer nature' (人定胜天). As Judith Shapiro (2001, 3–4) describes, official Maoist discourse was filled with military metaphors: 'Nature was to be "conquered". Wheat was to be sown by "shock attack". "Shock troops" reclaimed the grasslands. "Victories" were won against flood and drought. Insects, rodents, and sparrows were "wiped out"'.

Although Maoist China did not participate in practices of international society, it was not entirely isolated from the international community. Jin Canrong (2011, 8) describes the PRC's alliance with the Soviet Union as the fulfilment of China's 'important limited responsibility' and, to some extent, its 'leadership responsibility'. Accordingly, in its 'responsibility of maintaining the survival, unity and development of socialist countries in the world', China supported the socialist bloc during the Korean War, the First Indochina War and the Vietnam War. Once China and the Soviet Union abandoned their alliance, however, China began to expand its relations with former colonial states around the world. In its 'third-World' relations, China promoted international norms and values that were 'anti-America, anti-Soviet Union, [and] anti-colonialism' (ibid., 9). Despite Maoist China's impoverished state, it was 'somehow too generous' in providing its fellow communist countries with foreign aid, which prevented the party-state from fulfilling its national responsibility to raise the standard of living of its generally poor citizenry.[2] In return for China's diplomatic and economic assistance, many developing countries supported the PRC's bid to become China's legal representative at the UN. The PRC's membership in the UN and particularly its permanent seat on the UN Security Council re-oriented the state's international responsibilities, at least in principle, in the early 1970s. In reality, however, China remained unwilling to engage in all practices of international society, including those related to international trade, until the reform era. Nevertheless, because the UN Conference on the Human Environment (1972) was the first international conference in which the PRC ever participated, ideas of environmental responsibility have informed the PRC's formulation of its international responsibilities since remarkably early stages.

Despite substantial ideological and economic changes in China since the beginning of the post-Maoist era, Mao's views on foreign affairs and human–nature relationships continue to heavily influence China's contemporary environmental practices, and many of its current environmental problems can be traced to the Maoist era. Not only do China's development policies continue to focus on economic growth and largely neglect the environment, but the rule of law is weak, and public awareness of environmental problems remains low. In contrast to the Western focus on protecting nature and restoring native ecosystems, Chinese environmental projects, which aim to construct and improve nature by planting trees and implementing engineering innovations, have not managed to halt environmental degradation (Jiang 2010).

Reforming responsibilities

China's reform era began after Mao's death in the late 1970s. In their efforts to cultivate a moderately well-off society, Chinese leaders conceived economic growth primarily in terms of increased gross domestic product and income per capita. At the beginning of the reform period, the Chinese government concentrated on domestic economic growth and remained quite passive in the global arena. The chief motive of Chinese foreign policy was to channel foreign investments and technology into China in order to boost the state's economic growth. Amid reforms and open-door policies, however, China's national identity gradually changed during the 1980s and 1990s. Although that shift was mostly motivated by economic interests, it nevertheless transformed China's membership in international society. Since the late 1990s, the Chinese government has been more active in the international arena and adopted a 'going out' (走出去) strategy as a guideline for Chinese foreign policy. Recognising that the material focus on economic development was causing enormous environmental degradation and social problems, Chinese leaders began to acknowledge the importance of environmental protection, and in the late 1980s, China developed numerous organisational, educational and scientific programmes focused on the environment (cf. Harrington 2005, 108).

In 1997, the 15th National Congress of the Chinese Communist Party (CCP) set the construction of an 'all-around well-off society' (小康社会) as the party's top priority (Jiang 1997). To secure a favourable international environment for economic development, China abandoned ideologist campaigns and adopted a more pragmatic approach to foreign policy. Advised by Premier Deng Xiaoping, the Chinese government kept a low international profile, developed a concept of international responsibility 'somewhat strange to foreigners' and ranked raising its citizens' living standards as the 'fundamental national responsibility' and the 'main way of defending the superiority of the socialist approach' (Jin 2011, 9). Accordingly, CCP leaders argued that fulfilling China's domestic responsibilities *was* a measure of taking international responsibility because the collapse of China would undoubtedly cause instability around the world. That perspective remains central to China's understanding of its global responsibility. In Jin's words, because 'it is a great responsibility for China to get its own things done', the 'internal requirement for China's international responsibility is to develop and enhance its own power' (ibid., 8). In light of China's status as a developing country, Zhao Qizheng (2012, 197), a former minister of the State Council Information Office of China, even maintained that the Chinese state's 'first and foremost responsibility is to develop its economy to give the Chinese people a better life'. A 2011 white paper on China's peaceful development echoed a similar idea: 'For China, the most populous developing country, to run itself well is the most important fulfillment of its international responsibility' (Information Office of the State Council of the People's Republic of China 2011).

National responsibility

International relations theory tends to assume that states have universally generalised national interests. For example, realists regard physical survival and security as the chief, if not sole, national interest of all states, which leaves little room for international responsibility. By some contrast, rationalists stress economic interests, which also rarely support international responsibility. The constructivist perspective, however, stresses that national interests are ultimately only ideas. Similarly, the English School suggests that each state has nation-specific interests that are precisely what the nation considers them to be (cf. Bull 2002 [1977], 63–64). That approach underscores that each state is not merely an actor but 'foremost a *space* of positions' (Pouliot & Mérand 2013, 36). As any state, China is not a unitary agent but a set of complex social relations with multiple competing interests. For instance, the ideas and interests promoted by the Environmental Ministry might differ significantly from those of, say, the Ministry of Commerce. Thus, China's national interests are not givens but reflect the outcomes of contingent political struggles over values and preferences among various factions of the CCP and state organs, which are somewhat influenced by non-state actors such as corporations, non-governmental organisations and the general public.

In today's global era, states have increasingly more interests in common, and the promotion of a state's national interests therefore does not automatically threaten or undermine the interests and needs of other states. Some national interests are quite stable, some change over time and some are merely matters of choice. At the same time, some interests reflect a state's social circumstances: its identity, culture, traditions, history and political system, among other factors. Because national interests are social constructions, they shift over time; therefore, interests that imperial or Maoist China considered to be vital do not necessarily comprise the national interests of China's current leadership. Although China's contemporary political practices are largely determined by the objectives of the CCP, they are also shaped by the nation's long cultural history. In general, the geopolitical conditions of states vary as well and exert permanent influence on their interests and goals regardless of the state's political and social practices. Consequently, each government of China, whether socialist or democratic, has been or will be concerned with how to feed its enormous population, whereas each government of a small island state will be concerned about rising sea levels caused by climate change. Inevitably, other goals are simply determined by human choices. To respond to external and internal events (e.g. crises), decision makers have to choose from among multiple courses of action, which naturally involves an ethical comparison of appropriate options available. In 1989, China's leaders decided to respond with guns and tanks to student demonstrations at Tiananmen Square, whereas the government opted to react to protests addressing sovereignty in Hong Kong in 2014 in a more tolerant way. In China, the influence of individual decision makers is also considerable, because the state's leadership is embodied in the chairperson of the CCP, whose personal experiences, values, ideologies and goals exert a distinct influence on China's overall objectives.

Unity and stability have always been the major responsibilities of Chinese rulers. Since the beginning of the reform era, economic growth has been the principle means of maintaining stability and has thus been an overriding priority of the Chinese party-state. As part of that priority, China's growing demand for energy and other natural resources has played a decisive role in the state's diplomacy. Not only has China begun to entrench its cooperation with resource-laden developing countries, but Chinese companies have also started to invest overseas. Among China's other core interests today are state sovereignty, the safeguarding of its basic systems and national security, territorial integrity and national reunification, overall social stability and sustainable economic and social development.[3] The country's foreign policy has therefore had to support and advance all of those interests. In the early 2000s, energy security emerged as a central national concern, for China requires increasingly vast quantities of energy to maintain its rapid economic growth largely based on fuel-intensive heavy industry. To sustain economic growth in today's globalised world, China uses diplomatic tools to gain access to markets, foreign investment, advanced technology and energy and other natural resources.[4]

Since President Xi Jinping took office in March 2013, China has defended its national interests more vociferously in international politics. In his inauguration speech, Xi also introduced his vision of the *China Dream* (中国梦): 'Realizing the great renewal of the Chinese nation is the greatest dream for the Chinese nation in modern history' (quoted in Xinhua 2012a). In contrast to the American dream, which prizes individual success, the China dream seems to value national glory. Immediately after the speech, scholars and internet users began to discuss the definition and content of the concept. For example, Zheng Bijian, former executive vice president of the Party School of the CPC Central Committee and founding father of the concept of peaceful rise, suggested, 'Fundamentally, the Chinese dream is about using a peaceful, civilized way to realize national development and the modernization of a socialist country.' Similarly, Chinese scholar Yi Zhongtia explained that the Chinese dream prioritises 'national prosperity, social progress and people's happiness', as quotations in Xinhua (2012a) demonstrate. The China dream quickly emerged as the cornerstone of the Xi administration and can now be interpreted as the state's 'grand strategy'. In particular, the strategy draws from two 'centenary goals' to be achieved by the centennial of the PRC's founding in 2049: 'doubling the 2010 GDP and per capita income of urban and rural residents and finishing the building of a society of initial prosperity in all respects' and 'turning China into a modern socialist country that is prosperous, strong, democratic, culturally advanced and harmonious' (Xinhua 2014a). Although both goals had already been introduced in 1997 by then-President Jiang Zemin at the 15th National Congress, Xi Jinping's China dream has elevated the goals to the status of strategic national priorities for China. In practice, such elevation has positioned economic development and building a harmonious society as China's key priorities, even if the official ultimate goal of the government continues to be the realisation of communism.

Compared to Western legal systems, the Chinese system demonstrates significantly different ideas about responsibility. For one, whereas Western systems, generally based upon democracy and the rule of law, emphasise that the government is 'responsible to the people', the Chinese paradigm conceives the government to be 'responsible *for* the people' (Dobson 2013, 63). As Jiang Zemin (1997) puts it, the CCP 'shoulders a lofty historical responsibility for the destiny of the Chinese nation'. By putting development as a top priority, the CCP's constitution articulates the party's responsibilities for the people as follows:

> The general starting point and criterion for judging all the Party's work should be how it benefits development of the productive forces in China's socialist society, adds to the overall strength of socialist China and improves the people's living standards.
>
> (Xinhua 2012b)

By extension, in his inaugural press conference, President Xi (2012) similarly defined his responsibilities as a leader for the party, the nation and the Chinese people, partly in response to events from China's recent past. Following

China's economic reform and the CCP's legitimacy crisis after the Tiananmen Square incident in 1989, the Maoist moral basis of the CCP gradually crumbled. To legitimate its authority, the CCP has therefore been forced to rewrite its moral guidelines and, in doing so, has increasingly appropriated ideas from traditional Chinese schools of thought, especially Confucianism.[5] That shift has transformed not only the party itself but also its identity. Since 2012, Xi has accordingly led a national campaign to establish an all-pervasive 'moral and ideological foundation' for the country. According to Xi, socialist core values, including 'prosperity, democracy, civility, harmony, freedom, equality, justice, the rule of law, patriotism, dedication, integrity and friendship', should guide all public and private life in 'socialist China' (Xinhua 2014b).

International responsibility

China's rising economic wealth and gradual shift in identity have spawned debate over the state's international responsibilities since the 1990s (Xia 2001). As China's leaders began to consider that international organisations could benefit their nation's development, they did not want to portray China as a threat and thus started to present a global image of China as a responsible major power (Deng 2008; Gries 2004; Johnston 1998). Since then, two overarching beliefs have shaped China's political practices.[6] On the one hand, it is widely believed that China's future is increasingly linked to international society (Medeiros 2009, 20). The Chinese government recognises that China cannot develop in isolation and that its economic growth and national revitalisation depend heavily upon globalisation and other international practices. A stable internal and external security environment is similarly important for China's continued economic growth. The government has thus abandoned Maoist scepticism of international cooperation, and to date, China has signed more than 400 multilateral and more than 23,000 bilateral treaties as well as secured membership in nearly all international organisations (Liu 2015). It has even established new regional multilateral groups such as ASEAN Plus Three and the Shanghai Cooperation Organization. Without a doubt, China therefore seeks to fulfil its international responsibility to cooperate. On the other hand, perceptions of various internal and external threats, largely informed by communism, continue to guide China's political practices (Medeiros 2009, 20). Among the most pervasive beliefs in China is that the United States seeks to constrain China's development (ibid., 30). Younger Chinese scholars of international relations explain that belief with reference to a realist paradigm, whereas older ones often emphasise the Maoist class struggle against the United States, which they consider to be 'bellicose, unpredictable, and determined to hem China in' (Dobson 2013, 94). In either case, the belief partly explains why China has not accepted the US discourse of great power responsibility and, somewhat consequently, why China opposes legally binding greenhouse gas emission reductions.

Although formulated in 1954, the five principles of peaceful coexistence – mutual respect for sovereignty and territorial integrity, mutual non-aggression, non-interference in the internal affairs of other states, equality and mutual benefit and peaceful coexistence – persist as the cornerstone of China's practices of international responsibility (Xi 2014). All five principles are highly pluralist and state-centric in nature. Since 2004, the official foreign policy of China as derived from the concept of peaceful development (和平发展), which assures observers that a rising China does not seek hegemony and does not intend to challenge the current world order. The concept also stresses that the government's primary interest is promoting economic development and that China remains uninterested in interfering in conflicts as well as supports a stable international order. Instead of the traditional thinking of zero-sum security informed by the Cold War and based upon military force, the Chinese government now emphasises the importance of win–win security and progress in international politics. It moreover believes that international cooperation needs to be based on mutual trust and mutual benefit so that all involved parties win. That win–win rhetoric also suggests that China's economic growth will not threaten other states but instead offer them outstanding business opportunities. Extended to climate policy, it also assures the world that the Chinese government does not pose an economic, political or environmental threat to other states but instead unremittingly works to achieve harmony and welfare around the world.

When the Chinese government launched the concept of a 'harmonious world' (和谐世界) at the 2005 UN Summit, it was likely no coincidence that then-president Hu Jintao chose to do so at what was the UN's 60th anniversary summit. In traditional Chinese culture, 60 years marks the natural cycle of the end of one era and the birth of another. The concept of harmonious world stressed three diplomatic strategic viewpoints, all of which emphasise the maintenance of friendly relations with all countries, mutual respect and mutual understanding in China, 'non-enemy diplomacy', the construction of stable regional cooperation and global inter-regional cooperation (Su 2009, 54–55). Theoretically, the concept of harmonious world has attracted considerable attention among observers of China, who view it as a 'result of the rise of China's international position' or as a 'change of Chinese attitudes towards the outside world' and an 'intention to be a "responsible power"' (Masuda 2009, 58–59). In world politics, however, the concept has remained blurred and uninfluential. Consequently, Pang Zhongying (2006, 10) proposes that the lexicon of harmonious world should be 'translated from the realm of "good wishes" into formulating policy recommendations, identifying the challenges requiring China to assume a greater leadership role and developing the norms, rules and institutions that will define the international order of the 21st century'.

After Robert Zoellick (2005) famously urged China to become a responsible stakeholder, the global financial crisis of 2007/8 also elevated China's international status, whether or not the Chinese government desired or was ready for such a status. By extension, when C. Fred Bergsten (2008) complained that 'China's international mindset has not kept pace with its breathtaking economic ascent', he recommended the formulation of the G2, a 'true partnership' between China and the United States to 'provide joint leadership' for global economic practices. Similar written observations characterised China as the 'pioneer' of a new international order, which frightened most Chinese citizens (Jin 2011, 13). All of those instances aptly illustrate how participants in international practices assign roles and responsibilities to other participants, who in turn may feel pressured to assume more responsibilities than they are willing to or can shoulder. From China's perspective, its global responsibilities need to suit its level of development, and for now, China needs to primarily focus on promoting its development.

Nevertheless, the Chinese government is aware that its permanent seat at the UN Security Council comes with a special responsibility to maintain international peace and security (Wang 2013). Furthermore, the Chinese generally agree that China's rise indeed comes with farther-reaching responsibilities, and they currently debate 'about what responsibilities to assume, how to fulfil them, and how to balance between China's own abilities and other's expectations' (Jin 2011, 253). At the same time, they ask what sort of 'corresponding rights' China can attain, if any, should it assume greater international responsibilities (ibid., 18). According to Zhao (2013, 53), the debate among Chinese academics and policymakers has culminated in three major viewpoints on China's responsibility. The first recommends abandoning such a low-profile policy and victimised identity and instead assuming 'great power responsibility' in a bid to 'ensure a "just" world order' (ibid.). Conversely, the second viewpoint encourages China to take a 'more active, or even a leadership role' in order to promote China's core interests. Last, the third viewpoint supports China's continuation of a low-profile policy and thus advises avoiding more global responsibilities (ibid.). Meanwhile, the Chinese government has not remarked upon what sort of global power it should be but remains concerned about 'whether and how China can rise peacefully' (Tang 2006, 130).

Humanitarian responsibility

Although China's initial priority following its reform in 1978 was to 'get rich first', the Hu Jintao–Wen Jiabao administration (2003–2013) observed that such rapid economic growth had caused various environmental and social problems. Consequently, China's leaders began to pay more attention to cultivating social justice and reducing poverty as well as introduced 'putting people first' (以人为本) as the core principle of the social contract between the Chinese people and their political leaders. The principle maintains that China's leaders have to implement policies that benefit people and should not seek their own interests. As President Hu (2007) explained: 'We must always put people first. Serving the people wholeheartedly is the fundamental purpose of the Party, and its every endeavor is for the well-being of the people.'

The Hu-Wen administration coined two ideological concepts, both of which operationalise the principle of putting people first. The first one, the 'scientific outlook on development' (科学发展观) 'takes development as its essence, putting people first as its core, comprehensive, balanced and sustainable development as its basic requirement, and overall consideration as its fundamental approach' (Hu 2007). The concept also 'provides new scientific answers to the major questions of what sort of development China should achieve in a new environment and how the country should achieve it' (Hu 2012a). According to Hu, the concept was the 'most important achievement' of his leadership because it introduced a 'new realm in the development of Marxism in contemporary China' by 'integrating Marxism with the reality of contemporary China and with the underlying features of our times' (ibid.). At the 18th CCP National Congress in 2012, the scientific outlook on development was added to the revised Constitution of the CCP, which calls the concept 'a scientific theory' that 'puts people first and calls for comprehensive, balanced and sustainable development' (Xinhua 2012b). In addition, the 18th CCP National Congress affirmed the concept of 'protecting nationals abroad' (海外公民保护) as its priority, thereby reflecting the state's gradual acceptance of its national responsibility to defend its citizens overseas (Parello-Plesner & Duchâtel 2015). That responsibility was not, however, accepted in light of China's identity as a great power but due to its economic interests in unstable states.

The second ideological concept of the Hu-Wen administration was that of the harmonious society (和谐社会), which emphasises that '[s]ocial harmony is an essential attribute of socialism with Chinese characteristics' (Hu 2007). The Sixth Plenum of the 16th CCP Central Committee declared in 2006 that 'social harmony is the intrinsic nature of the socialism with Chinese characteristics and an important guarantee of the country's prosperity, the nation's rejuvenation and the people's happiness' (People's Daily 2006a). At the 17th Party Congress in 2007, the concept of a harmonious society was institutionalised and added to the CCP's constitution, thereby signalling that the CCP was 'formally giving up Maoist "class struggle"' (Brady 2012, 66). Likewise, Hu's report to the Party Congress abandoned calls for 'establishing a "new international political and economic order", a phrase that Deng started using in these reports as far back as 1988' (Medeiros 2009, 49). Instead of challenging the practices of international society, contemporary China has played a 'constructive role' and worked to 'make the international order fairer and more equitable' (Hu 2007).

In essence, the concepts of scientific development and social harmony are 'two sides of the same coin: a harmonious society is the objective and scientific development is the method to reach it' (Fan 2006, 709). According to Hu (2007), 'scientific development and social harmony are integral to each other and neither is possible without the other'. Both concepts can be understood as means to and incentives for sustainable development, while putting people first does not necessarily mean that the Chinese government will ignore the environment. On the contrary, the government recognises that environmental degradation obstructs sustainable economic growth and that pollution harms the wellbeing of people. Scientific development and social harmony therefore provide China's government with strong anthropocentric incentives for environmental protection. Recently, China has begun to heed citizens' environmental rights as well. A white paper addressing 'Progress in China's Human Rights in 2014' dedicated more than an entire chapter to environmental protection for the first time. The chapter reviews China's governmental policies and actions for safeguarding citizens' environmental rights to live in clean, healthy environment and have a good eco-environment, as well as their interests in achieving those goals (Information Office of the State Council of the People's Republic of China 2013). Environmental protection has indeed become an important part of China's national policy, as I discuss at length in the next section.

Although China has ratified numerous humanitarian treaties, including all four Geneva Conventions and the Universal Declaration of Human Rights, its approach to human rights differs from that of Western nations. In short, the Chinese regard human rights as a domestic affair, and such rights are therefore a state's national responsibility, not a humanitarian one. However, such thinking does not necessarily exclude the concept of humanitarian responsibility from the Chinese context. Indeed, China has increasingly supported international humanitarian assistance (cf. Chan 2006, 37), no longer believes that UN peacekeeping operations are 'interference in countries' internal affairs' and the 'undesirable result of US–Soviet hegemonic power competition' (Wang 1999, 75) and has more actively engaged in UN peacekeeping operations since 1990. However, developed countries often complain that China's growing presence in Africa, for instance, negatively affects democratisation, human rights, good governance and environmental protection there. Of course, China denies those claims and assures that though it emphasises 'both morality and interests' in its exchanges with other developing countries, it prioritises 'morality before interests' (Wang 2013).

Environmental responsibility

Although environmental protection lacked its own chapter in China's five-year plans before the sixth Five-Year Plan (1981–1985), China established its first environmental regulations in 1973 (Ross 1999, 298–299). That same year, China also held the first national conference on environmental protection, where a 'firm decision was taken on the need for environmental safekeeping' (Palmer 1998, 790). A year later, China established the small Environmental Protection Office under the State Council (Ross 1999, 298), and in 1978, it added environmental protection to the PRC's constitution. According to Article 11 of that constitution, 'The state protects the environment and natural resources, and prevents and eliminates pollution and other hazards to the public' (Palmer 1998, 791). China also passed the Environmental Protection Law for trial use in 1979. Political and economic reforms catalysed environmental policies, and during the 1980s, many new environmental laws were enacted.[7] The constitutional promise of environmental responsibility was expanded in Article 26 of the 1982 constitution, which declares, 'The State protects and improves the environment in which people live and the ecological environment. It prevents and controls pollution and other public hazards (Database for Laws and Regulations)'. Interestingly, the article differentiates 'the environment in which people live' and 'the ecological environment', which suggests that humans are not an integral part of the latter.

President Jiang Zemin's report to the 15th Party Congress held in 1997 was the first-ever report of the National Congress to mention the environment. According to Jiang (1997), 'While exploiting our [China's] natural resources and making economical use of them, we lay emphasis on the latter so as to raise the efficiency of their utilization.' Jiang particularly acknowledged that 'population growth and economic development have caused great strains on resources and the environment' and that environmental problems could hinder China's development in the future (ibid.). Given previous beliefs that only capitalism caused environmental degradation, the acknowledgement marked a major shift in the PRC's discourse.

At the 16th Party Congress, President Jiang (2002) pointed out that the 'contradiction between the ecological environment and natural resources on the one hand and economic and social development on the other is becoming increasingly conspicuous'. In 2004, China launched a highly publicised campaign on so-called 'green GDP' in order to integrate environmental protection into economic practices (Economy 2007, 27–30). It pursued the development of an index that 'quantifies and measures the monetary costs of environmental damage caused by a country's economic growth' (Zheng 2015). Due to technical difficulties, however, the world's first green GDP report, published in 2006, calculated only environmental pollution's economic costs (People's Daily 2006b). Subsequently, the green GDP campaign was quickly abandoned due to methodological problems and the overriding priority of economic goals at the regional level.

Both before and especially after the introduction of the harmonious society concept, many Chinese intellectuals began to call for ecological civilization as a new model for achieving harmony between nature and humanity (Dynon 2008). As Ma Jun (2007) puts it, 'If the aim of development is really to benefit the people, we cannot destroy the very resources on which people rely for survival.' Similarly, Pan Yue (2006), vice-minister of the State Environmental Protection Administration, stated in his two influential articles 'Harmonious Society and Environment-Friendly Society' and 'On Socialist Ecological Civilization' that harmonious society cannot be achieved without environmental protection. He proposed that harmonious society should be

> based on industry consuming less resources, people's livelihoods based on moderate-level consumption, greater recycling of resources, a highly efficient economic infrastructure, greater innovation, an orderly and open financial set-up, a distribution system emphasizing social justice and a democratic political system.
>
> (Pan 2006)

President Hu responded to such calls by presenting new requirements for developing a moderately well-off society in his report to the 17th CCP National Congress in 2007, as well as by officially proposing the development of an ecological civilisation (生态文明) for the first time.[8] According to Hu (2007), China should pursue ecological civilisation 'by basically forming an energy- and resource-efficient and environment-friendly structure of industries, pattern of growth and mode of consumption'. Although Hu did not clearly define *ecological civilisation*, his report indicated that the government had redefined its development model by increasing emphasis on sustainable development. The concept of ecological civilisation was quickly incorporated into the government's overall policy plans and added to the CCP's constitution in 2012 (cf. Xinhua 2012b). At the 18th National Congress of the CCP in 2012, the concept received its own chapter in the conference report for the first time, and Hu mentioned the concept of ecological civilisation 15 times in the report, up from only twice in the 2007 report. In the more recent report, Hu (2012b) acknowledged that promoting an ecological civilisation 'is a long-term task of vital importance to the people's wellbeing and China's future' and thus deserving of 'high priority'. He declared that the CCP would incorporate the concept 'into all aspects and the whole process of advancing economic, political, cultural, and social progress, work hard to build a beautiful country, and achieve the lasting and sustainable development of the Chinese nation' (ibid.). Moreover, delegates to the 18th Party Congress noted that China had entered a 'new stage of development', meaning that it should no longer focus exclusively on rapid economic growth but also integrate environmental protection, emissions reduction and energy conservation into its overall development targets. However, economic growth has persisted as a precondition of environmental protection. As China's minister of environmental protection, Zhou Shengxian, puts it: 'We [China] must make sure that environmental protection is an essential part of the efforts to promote economic growth. It is impossible to achieve environmental protection without economic growth, because this would be like catching fish on a tree' (quoted in CCTV 2012).

Since the outset of the Xi Jinping–Li Keqiang administration in 2013, the development of an ecological civilisation has received wide attention from CCP elites. In the spring of 2015, the State Council effectively elevated the concept to the status of a prominent strategic guideline to be integrated into China's economic, political, cultural and societal plans. In addition to the transformation of economic development, 'Opinion on Acceleration for the Promotion of Ecological Civilisation' explained 'accelerating the promotion of ecological civilisation' to mean

> improving internal requirements regarding the quality and efficiency of development, promoting the 'putting people first' principle and the inevitable choice of social harmony, fostering a moderately well-off society, realising the great rejuvenation of the Chinese dream,

> responding actively to climate change and safeguarding global environmental security.
>
> (Xinhua 2015, my translation)

Xi's (2017, 47) report to the 19th CCP Congress in October 2017 placed great stress on environmental issues by not only promoting green development and the construction of ecological civilisation but also urging China to 'do our generation's share to protect the environment'. Surprisingly, he also expressly advocated the importance of environmental conservation:

> Man and nature form a community of life … . Only by observing the laws of nature can mankind avoid costly blunders in its exploitation. Any harm we inflict on nature will eventually return to haunt us. This is a reality we have to face.
>
> (Ibid.)

Ample literature has elaborated upon the concept of ecological civilisation in China.[9] Whereas Western scholars tend to understand the concept to be more or less synonymous with sustainable development and to focus on greener economic development and environmental protection, Chinese ones understand ecological civilisation as the 'level of harmony that exists between human progress and natural existence in human civilization' (Ma 2013). They thus emphasise the cultural and socialist dimensions of the concept and regard it as the stage of social progress following 'the primitive civilization, the agricultural civilization and the industrial civilization' (Pan 2006). Moreover, they posit, the concept addresses the weaknesses of the most recent stage of (Western) industrial civilisation and aims to strike a better balance between the environment and development. The label *civilisation* affords environmental protection a higher position in the CCP's hierarchy of values, thereby indicating environmental protection is, at least in principle, evolving into a moral or ideological imperative (Dynon 2008, 107). Ecological civilisation can be thus grouped with other post-Maoist morality campaigns that seek to formulate what it means to be civilized. The three other concepts of the CCP's civilisation narrative – material, spiritual and political civilisation – have played important roles in establishing and maintaining the party's moral legitimacy (ibid.).

Many Chinese intellectuals have encouraged the CCP to develop the concept of ecological civilisation by adding ideas from Marxist ecology and traditional Chinese religion and philosophy (cf. Wang 2012; Wang, He & Fan 2014; Pan 2006). In particular, some Chinese Marxist theorists have suggested that Chinese scholars develop their own Chinese ecological Marxist theory and not 'treat ecological Marxism as a foreign dogma to be worshipped but a living method with which to analyze and solve the serious [environmental] problems facing China' (Wang 2012). Unlike previous civilisations, which were largely motivated by domestic development and targeted a domestic audience,[10] an ecological civilisation could have far-reaching global impacts both in theory and practice. Given the shortcomings of the concept of sustainable development, the concept of ecological civilisation could also provide food for thought for more solidarist international practices, particularly in representing a new, holistic worldview unlike anthropocentrism by viewing humans as the core but not the masters of nature. Moreover, it is a concept that 'knows no boundaries' but acknowledges that the 'balance between humans and nature must be approached on a global basis' (Ma 2013). However, other Chinese scholars doubt the global possibilities of the concept because they regard socialism as a precondition of ecological civilisation. Because the Western world has failed to develop an ecological civilisation, Pan (2006) believes that China should spearhead the development of a new ecological model of life. For the time being, however, the idea of ecological civilisation has not been translated into an ecocentric policy practice. For example, China's nationally determined contribution to the United Nations Framework Convention on Climate Change in June 2015, which I discuss in chapter 6, announced that China would 'work together with all Parties to build a beautiful homeland for all *human beings*' (National Development and Reform Commission 2015, 16, italics added) but ignores the intrinsic value of the environment.

Conclusion

In this chapter I have investigated historical notions of responsibility in China. In focusing exclusively on official historical narratives that China's political elite has pursued to legitimate, maintain and strengthen its position in the one-party system, I have dismissed alternative approaches to state responsibility despite their potential to provide more solidarist ideas of responsibility. Had I studied social processes in which definitions of *responsibility* and its allocation – processes that I refer to collectively as *responsibilisation* – in greater detail, I would have paid more attention to questions such as who has been able to participate in processes of defining notions of responsibility in China and who has been marginalised from those processes at different times. I would have also sought to identify what sort of social, political and economic ramifications such inclusion and exclusion have caused domestically and how they have guided China's international role. At the same time, it would have been important to study notions of responsibility in Chinese philosophy and political schools of thought in order to pinpoint which philosophical ideas of responsibility have shaped Chinese notions of responsibility and whether they could provide the world with a more solidarist basis for organising international society. For the purposes of this book, however, a more elaborate study would be excessive to maintaining a state-centric focus.

Notes

1 That view comes close to what Dryzek (2005) calls *Promethean* or *cornucopian* discourse.
2 According to Chen Z. (2009, 25), foreign aid expenditure accounted for 6.9% of the PRC's state budget in 1973.
3 According to China News Agency (2009), then-State Councillor Dai Bingguo identified China's core interests at the first China–US Strategic and Economic Dialogue.
4 For more detailed accounts and analyses of Chinese foreign policy, see Lanteigne (2013) and Liu (2004).
5 See Brady (2012) for the evolution of Chinese state Confucianism, from so-called Maoist 'smash Confucianism' via Chinese studies fever in the 1990s to the sophisticated use of Confucian terms in CCP ideology in the 2000s.
6 Although Medeiros's (2009) argument of the two beliefs refers only to China's views of international security, I argue that they also influence China's role in international society in general.
7 About the development of Chinese environmental law, see Chen G. (2009) and Palmer (1998).
8 Although the concept of ecological civilisation was initially translated as 'conservation culture' or 'ecological progress' in official documents, the term *ecological civilisation* was quickly standardised as the official translation.
9 Western observers of China, however, have only recently become interested in ecological civilisation. Even the 2013 *China Story Yearbook*, titled 'Civilising China', mentioned ecological civilisation in only one brief sentence, which indicates that Western scholars do not consider that the concept has found its ultimate place in the CCP toolbox.
10 However, see Nyiri (2006) for an elaboration upon the so-called 'yellow man's burden' to civilise developing countries.

Bibliography

Bergsten, C. Fred. 2008. 'A partnership of equals'. *Foreign Affairs* 87:4, 57–69.

Brady, Anne-Marie. 2012. 'State Confucianism, Chineseness, and tradition in CCP propaganda'. In Brady, Anne-Marie (ed.), *China's Thought Management*. New York: Routledge, 57–75.

Bull, Hedley. 2002 [1977]. *The Anarchical Society: A Study of Order in World Politics*, 3rd edition. Basingstoke: MacMillan Press.

CCTV. 2012. 'Preserving the environment a major concern for China', 13 November. Accessed 3 March 2017. http://english.cntv.cn/program/china24/20121113/107731.shtml.

Chan, Gerald. 2006. *China's Compliance in Global Affairs: Trade, Arms Control, Environmental Protection, Human Rights*. Singapore: World Scientific Publishing Co.

Chen, Gang. 2009. *Politics of China's Environmental Protection: Problems and Progress*. Singapore: World Scientific Publishing.

Chen, Zhimin. 2009. 'International responsibility and China's foreign policy'. In Iida Masafumi (ed.), *China's Shift: Global Strategy of the Rising Power*. NIDS Joint Research Series no 3. Tokyo: National Institute for Defense Studies, 7–28.

China News Agency. 2009. '首轮中美经济对话:除上月球外主要问题均已谈及' [The first round of Sino-US economic dialogue discusses all the important issues], 29 July. Accessed 3 March 2017. www.chinanews.com/gn/news/2009/07-29/1794984.shtml

Database for Laws and Regulations. Accessed 3 March 2017. http://www.npc.gov.cn/englishnpc/Law/2007-12/05/content_1381903.htm.

Deng, Yong. 2008. *China's Struggle for Status: The Realignment of International Relations*. Cambridge: Cambridge University Press.

Dobson, Wendy. 2013. *Partners and Rivals: The Uneasy Future of China's Relationship with the United States*. Toronto: University of Toronto Press.

Dryzek, John S. 2005. *The Politics of the Earth: Environmental Discourses*. New York: Oxford University Press.

Dynon, Nicholas. 2008. '"Four civilizations" and the evolution of post-Mao Chinese socialist ideology'. *China Journal* 60, 83–109.

Economy, Elizabeth. 2007. 'Environmental governance: The emerging economic dimension'. In Neil T. Carter & Arthur P. J. Mol (eds), *Environmental Governance in China*. London: Routledge, 23–41.

Fan, Cindy C. 2006. 'China's Eleventh Five-Year Plan (2006–2010): From "getting rich first" to "common prosperity"'. *Eurasian Geography and Economics* 47:6, 708–723.

Greenfield, Jeanette. 1979. *China and the Law of the Sea, Air, and Environment*. Alphen aan den Rijn: Sijthoff & Noordhoff.

Gries, Peter H. 2004. *China's New Nationalism: Pride, Politics, and Diplomacy*. Berkeley, CA: University of California Press.

Harrington, Jonathan. 2005. "Panda diplomacy': State environmentalism, international relations and Chinese foreign policy'. In Paul. G. Harris (ed.), *Confronting Environmental Change in East and Southeast Asia: Eco-Politics, Foreign Policy, and Sustainable Development*. London: United Nations University Press and Earthscan, 102–118.

Hu, Jintao. 2012a. 'Promote win-win cooperation and build a new type of relations between major countries'. Accessed 3 March 2017. www.fmprc.gov.cn/mfa_eng/wjdt_665385/zyjh_665391/t931392.shtml.

Hu, Jintao. 2012b. 'Firmly march on the path of socialism with Chinese characteristics and strive to complete the building of a moderately prosperious society in all respects'. Accessed 3 March 2017. www.china.org.cn/china/18th_cpc_congress/2012-11/16/content_27137540.htm.

Hu, Jintao. 2007. 'Hold high the great banner of socialism with Chinese characteristics and strive for new victories in building a moderately prosperous society in all respects'. Accessed 3 March 2017. http://www.china.org.cn/english/congress/229611.htm.

Information Office of the State Council of the People's Republic of China. 2013. 'Progress in China's human rights in 2012'. Accessed 3 March 2017. http://english.gov.cn/archive/white_paper/2014/08/23/content_281474982986492.htm.

Information Office of the State Council of the People's Republic of China. 2011. 'China's peaceful development'. Accessed 3 March 2017. www.gov.cn/english/official/2011-09/06/content_1941354.htm.

Jiang, Hong. 2010. 'Desertification in China'. In Joel Jay Kassiola & Sujian Guo (eds), *China's Environmental Crisis: Domestic and Global Political Impacts and Responses*. New York: Palgrave Macmillan, 13–40.

Jiang, Zemin. 2002. 'Full text of Jiang Zemin's Report at 16th Party Congress.' Accessed 3 March 2017. www.china.org.cn/english/features/49007.htm.

Jiang, Zemin. 1997. 'Hold high the great banner of Deng Xiaoping theory for an all-round advancement of the cause of building socialism with Chinese characteristics' into the 21st century'. Accessed 3 March 2017. www.bjreview.com.cn/document/txt/2011-03/25/content_363499.htm.

Jin, Canrong. 2011. *Big Power's Responsibility: China's Perspective.* Translated by TuXiliang. Beijing: China Renmin University Press.

Johnston, Alistair I. 1998: 'China and international institutions: A decision rule analysis'. In Michael B. McElroy, Chris P. Nielson & Peter Lydon (eds), *Energizing China: Reconciling Environmental Protection and Economic Growth.* Cambridge, MA: Harvard University Press, 555–599.

Kobayashi, Yuka. 2005. 'The "troubled modernizer": Three decades of Chinese environmental policy and diplomacy'. In Paul. G. Harris (ed.), *Confronting Environmental Change in East and Southeast Asia: Eco-Politics, Foreign Policy, and Sustainable Development.* London: United Nations University Press and Earthscan, 87–101.

Kubin, Wolfgang. 2010. 'The myriad things: Random thought on nature in China and the West'. In Hans Ulrich Vogel & Günter Dux (eds), *Concepts of Nature: A Chinese-European Cross-cultural Perspective.* Leiden: Brill, 516–525.

Lanteigne, Marc. 2013. *Chinese Foreign Policy*, 2nd edition. New York: Routledge.

Liu, Guoli (ed.). 2004. *Chinese Foreign Policy in Transition.* New York: Aldine De Gruyter.

Liu, Zhenmin. 2015. 'Uphold the authority of the UN Charter and promote win-win cooperation'. Accessed 3 March 2017. www.fmprc.gov.cn/mfa_eng/wjdt_665385/zyjh_665391/t1255145.shtml.

Ma, Jun. 2007. 'Ecological civilisation is the way forward'. *China Dialogue*, 31 October. Accessed 13 March 2017. www.chinadialogue.net/article/1440-Ecological-civilisation-is-the-way-forward.

Ma, Kai. 2013. 'Committing to the development of an ecological civilization'. *Qiushi Journal* 5:4. Accessed 3 March 2017. http://english.qstheory.cn/magazine/201304/201311/t20131107_288028.htm.

Marks, Robert B. 2012. *China: Its Environment and History.* Lanham: Rowman & Littlefield.

Masuda, Masayuki. 2009. 'China's search for a new foreign policy frontier: Concept and practice of "harmonious world"'. In Iida Masafumi (ed.), *China's Shift: Global Strategy of the Rising Power.* NIDS Joint Research Series no 3. Tokyo: National Institute for Defense Studies, 57–79.

Medeiros, Evan S. 2009. *China's International Behavior.* Santa Monica, CA: RAND Corporation.

National Development and Reform Commission. 2015. 'Enhanced actions on climate change: China's intended nationally determined contribution'. Accessed 23 March 2017. http://www4.unfccc.int/submissions/indc/Submission%20Pages/submissions.aspx.

Nyiri, Pal. 2006. 'The yellow man's burden: Chinese migrants on a civilizing mission'. *The China Journal* 56 (July).

Palmer, Michael. 1998. 'Environmental regulation in the People's Republic of China: The face of domestic law'. *China Quarterly* 156, 788–808.

Pan, Yue. 2006. 'Evolution of an ecological civilization'. *Beijing Review* 45. Accessed 9 March 2017. www.bjreview.cn/EN/06-45-e/point-1.htm.

Pang, Zhongying. 2006. 'China, my China'. *National Interest* 83 (Spring), 9–10.

Parello-Plesner, Jonas & Mathieu Duchâtel. 2015. 'How Chinese nationals abroad are transforming Beijing's foreign policy'. *East Asia Forum*, 16 June. Accessed 3 March 2017.

www.eastasiaforum.org/2015/06/16/how-chinese-nationals-abroad-are-transforming-beijings-foreign-policy.

People's Daily. 2006a. 'Communique of the Sixth Plenum of the 16th CPC Central Committee'. 12 October. Accessed 3 March 2017. http://en.people.cn/200610/12/eng20061012_310923.html.

People's Daily. 2006b. 'GDP takes on a green hue in new figures'. 8 September. Accessed 3 March 2017. http://en.people.cn/200609/08/eng20060908_300803.html.

Pouliot, Vincent & Frédéric Mérand. 2013. 'Bourdieu's concepts'. In Rebecca Adler-Nissen (ed.), *Bourdieu in International Relations: Rethinking Key Concepts in IR*. London: Routledge, 24–44.

Ross, Lester. 1999. 'China and environmental protection'. In Elizabeth Economy & Michel Oksenberg (eds), *China Joins the World: Progress and Prospects*. New York: Council on Foreign Relations Press, 296–325.

Schmidt-Glintzer, Helwig. 2010. 'On the relationship between man and nature in China'. In Hans Ulrich Vogel & Günter Dux (eds), *Concepts of Nature: A Chinese-European Cross-Cultural Perspective*. Leiden: Brill, 526–542.

Shapiro, Judith. 2001. *Mao's War Against Nature: Politics and the Environment in Revolutionary China*. New York: Cambridge University Press.

Sit, Victor F. S. 2010. *Chinese City and Urbanism: Evolution and Development*. Singapore: World Scientific Publishing.

Su, Hao. 2009. 'Harmonious world: The conceived international order in framework of China's foreign affairs'. In Iida Masafumi (ed.), *China's Shift: Global Strategy of the Rising Power*. NIDS Joint Research Series no 3. Tokyo: National Institute for Defense Studies, 29–55.

Tang, Shiping. 2006. 'Projecting China's foreign policy: Determining factors and scenarios'. In Jae Ho Chung (ed.), *Charting China's Future: Political, Social, and International Dimensions*. Lanham, MD: Rowman & Littlefield, 129–145.

Wang, Jianwei. 1999. 'Managing conflict: Chinese perspectives on multilateral diplomacy and collective security'. In Yong Deng & Fei-Ling Wang (eds), *In the Eyes of the Dragon: China Views the World*. Lanham, MD: Rowman & Littlefield.

Wang, Yi. 2013. 'Exploring the path of major-country diplomacy with Chinese characteristics'. Accessed 3 March 2017. www.fmprc.gov.cn/mfa_eng/wjb_663304/wjbz_663308/2461_663310/t1053908.shtml.

Wang, Zhihe. 2012. 'Ecological Marxism in China'. *Monthly Review* 63:9. Accessed 4 March 2017. http://monthlyreview.org/2012/02/01/ecological-marxism-in-china.

Wang, Zhihe, Huili He & Meijun Fan. 2014. 'The ecological civilization debate in China'. *Monthly Review* 66:6. Accessed 4 March 2017. http://monthlyreview.org/2014/11/01/the-ecological-civilization-debate-in-china.

Weller, Robert P. 2006. *Discovering Nature: Globalization and Environmental Culture in China and Taiwan*. Cambridge: Cambridge University Press.

Zhang, Yongjin. 2014. 'Curious and exotic encounters: Europeans as supplicants in the Chinese Imperium, 1513–1793'. In Shogo Suzuki, Yongjin Zhang & Joel Quirk (eds), *International Orders in the Early Modern World: Before the Rise of the West*. New York: Routledge, 55–75.

Zhao, Suisheng. 2013. 'Core interests and great power responsibilities: The evolving pattern of China's foreign policy'. In Xiaoming Huang & Robert G. Patman (eds), *China and the International System: Becoming a World Power*. New York: Routledge, 32–56.

Zhao, Suisheng. 2000. 'Chinese nationalism and its international orientations'. *Political Science Quarterly* 115:1, 1–33.

Zhao, Qizheng. 2012. *How China Communicates: Public Diplomacy in a Global Age*. Beijing: Foreign Language Press.

Zheng, Jinran. 2015. 'Watchdog restarts study on green GDP'. *China Daily*, 1 April. Accessed 4 March 2017. www.chinadaily.com.cn/china/2015-04/01/content_19965714.htm.

Zoellick, Robert B. 2005. 'Whither China: From membership to responsibility? Remarks to National Committee on US-China Relations'. Accessed 4 March 2017. http://2001-2009.state.gov/s/d/former/zoellick/rem/53682.htm.

Xi, Jinping. 2017. 'Secure a decisive victory in building a moderately prosperous society in all respects and strive for the great success of socialism with Chinese characteristics for a new era'. Accessed 14 February 2018. http://www.xinhuanet.com/english/download/Xi_Jinping's_report_at_19th_CPC_National_Congress.pdf.

Xi, Jinping. 2014. 'Carry forward the five principles of peaceful coexistence to build a better world through win-win cooperation'. Accessed 4 March 2017. www.fmprc.gov.cn/mfa_eng/wjdt_665385/zyjh_665391/t1170143.shtml.

Xi, Jinping. 2012. 'Remarks on the occasion of meeting with the Chinese and foreign press by members of the Standing Committee of the Political Bureau of the Eighteenth Central Committee of the Communist Party of China'. Accessed 4 March 2017. www.china.org.cn/china/18th_cpc_congress/2012-11/16/content_27130032.htm.

Xia, Liping. 2001. 'China: A responsible great power'. *Journal of Contemporary China* 10:26, 17–25.

Xinhua. 2015. '授权发布 中共中央 国务院关于加快推进生态文明建设的意见' [Authorized release: An opinion of the Chinese Communist Party Central Committee and State Council regarding acceleration of promotion of the construction of ecological civilization], 5 May. Accessed 4 March 2017. http://news.xinhuanet.com/politics/2015-05/05/c_1115187518.htm

Xinhua. 2014a. 'Xi eyes more enabling int'l environment for China's peaceful development'. 30 November. Accessed 22 October 2016. http://en.people.cn/n/2014/1130/c90883-8815967.html.

Xinhua. 2014b. 'Xi stresses core socialist values'. 25 February. Accessed 22 October 2016. www.chinadaily.com.cn/china/2014-02/26/content_17305164.htm.

Xinhua. 2012a. 'Xi pledges "great renewal of Chinese nation"'. 29 November. Accessed 22 October 2016. http://news.xinhuanet.com/english/china/2012-11/29/c_132008231.htm.

Xinhua. 2012b. 'Full text of Constitution of Communist Party of China'. 18 November. Accessed 22 October 2016. www.xinhuanet.com/english/special/2017-11/03/c_136725945.htm.

4 China's rise, climate change and great power responsibility

The English School assumes that great powers have special responsibilities in international society. Those responsibilities, however, are not givens but socially constructed expectations, developed both implicitly and explicitly, in the so-called 'great power club'. Since the end of World War II, respect for human rights and the principle of the responsibility to protect have constituted essential attributes of the responsibility of states in the great power club, or what I call *great power responsibility*. In the twenty-first century, however, China's rise to the status of great power may in turn transform the conceptualisation of great power responsibility. Among the situations that China and other great powers have to address today and in the future is climate change, which has become an increasingly alarming threat to international security and the wellbeing of humankind. In this chapter, I therefore ask to what extent we can assume that great powers should shoulder more responsibility than smaller powers for mitigating climate change. In pursuing a normative discussion about great power climate responsibility from the perspective of the English School and other modes of thinking, I first introduce the concept of the great power club, after which I discuss normative dimensions of great power responsibility from the perspective of the pluralist–solidarist debate within the English School. Empirical parts of the chapter study the practices of great power responsibility and elaborate upon whether those practices have paid sufficient attention to environmental stewardship and, if so, then how. Last, I examine the sort of requirements that the United States, as an established great power, has set for China's membership in the great power club and how China has responded to those expectations.

The great power club

The concept of great power is exceptionally vague. From the perspective of the English School, however, a clear definition of *great power* is not necessary to understand and study their role in international society. On the contrary, what matters is how a state is constituted in international practices (cf. Frost 2003, 86). For a state to be a great power, it needs to have certain material capabilities and, more importantly, be recognised as a member of the great power club. In emphasising social participation and mutual engagement, Étienne Wenger's (1998, 76–77) concept of community of practice elucidates what the English School means by *great power club.* Although the term *community* often connotes positive interactions and peaceful co-existence, Wenger's concept of community of practice does not offer an 'idealized view of what a community should be' but stresses that a community of practice nevertheless exists because its participants remain 'engaged in actions whose meanings they negotiate with one another' (ibid., 73). However, a shared practice does not require consensus regarding all the rules of practice but practices can involve competition, tensions and even violent conflicts among the participants. Such disagreements connect and engage the participants in complex ways and can even generate changes in the practices themselves. For classic English School theorists such as Bull and Wight, great power management inevitably stands as one of the primary institutions of international society. Holsti (2009), however, disagrees with that standpoint because great power management, with the exception of the Concert of Europe, does not fulfil his criteria for patterned practices. For Holsti (2009, 137), great power is a status, not an institution. Today, it remains difficult, if not impossible, to pinpoint the geopolitical centre of the great power club. Apart from the UN Security Council, in which not all emerging powers are members, no secondary institution of great powers exists, although the Group of 8 is a candidate to some extent.

The great power club is an exclusive social community; whereas some states are members, others remain outsiders. Of course, the club has no membership card, so to speak, and the qualifications for being an accepted member change over time; the conditions of membership are not written into international treaties but based on the social order continuously shaped by the social interactions of states. Such interactions generate informal criteria for a state's achieving and maintaining status as a member of the club, as well as define perceptions of what behaviour is appropriate for great powers. Newcomers to the club such as China have to learn to follow the club's rules or can attempt to alter them with their words and actions (Kopra 2016; Kopra 2018). Although any set of social rules has to be upheld to some degree in order to be effective, their occasional violation is not unusual. Indeed, if the violation of rules from time to time was impossible, then having them at all would be pointless.[1] According to Suzuki (2014, 637), states today need to fulfil two conditions if they wish to join the great power club. First, they have to enjoy substantial institutional privileges in international decision making, as China clearly does. Second, they have to 'be treated as a *social* equal' with other members of club. China's questionable fulfilment of the latter condition has cast the greatest doubt over its bid for membership in the great power club, which has frustrated China's leaders and motivated their efforts to persistently improve the state's image in the international society (Suzuki 2008).

From the standpoint of the English School, international norms and practices constrain international society because they establish social guidelines for and barriers to what behaviour is conceived to be acceptable and legitimate for states (Wheeler 2000, 4–5). Nevertheless, societal legitimacy remains an important, if not the most important, condition of membership in the great power club. According to Bull (2002 [1977], 221), 'Great powers can fulfil their managerial functions in international society only if these functions are accepted clearly enough by a large enough proportion of the society of states to command legitimacy.' Such legitimacy relates closely to the rights and responsibilities of great powers, which are 'recognised by others to have, and conceived by their own leaders and peoples to have, certain special rights and duties' (ibid., 196). At the same time, those rights and responsibilities cannot be formalised, much less articulated, by writing out the hegemonial rights of great powers, because anarchical international society rejects the idea of any hierarchical ordering of states whatsoever (ibid., 221). Consequently, the great power club is too indefinite and vague a community of practice to set formal rules about how great powers should act. As all virtues, the virtues of the great power club have resulted from historical practices and may change in the future. Because it therefore falls to the great powers to negotiate which circumstances and demands transform the rules of membership in the great power club, China's rise could significantly shape the rules of great power management in time.

At present, although great powers are generally thought to have an informal responsibility to cooperate and take other states' interests into account, they have no concrete, formal obligation to act in certain ways. Consequently, though the practices of great power management specify what great powers ought to do, they do not prescribe means of performing such actions. Instead, they condone several ways to take action provided that such actions do not meet with international criticism. To be seen as legitimate, great powers are bound to promote, or at least take into account, international justice and other international demands. Although other states do not expect perfect performance, legitimate great powers have to avoid behaviour that could cause international disorder and injustice; they have the right to mould international practices, but their freedom of action is limited by their responsibility (Bull 2002 [1977]). At the same time, though smaller states and non-state actors can lobby great powers and remind them of their global responsibilities, ultimately the great powers themselves collectively define the rules of practice befitting great powers.

The very concept of great power at the international level implies a balance of power as well as the existence of a great power club. After all, if there were only one dominant state, then that state would not be a great power, for it would be impossible to compare and rank the statuses of other states and to construct social identities. Nevertheless, many observers have characterised the post-Cold War international system as unipolar due to the hegemonic dominance of the United States.[2] In the English School, Ian Clark (2009a, 205) also asks, 'What then happens to international order if there *is* only one predominant state?' (cf. Bukovansky et al. 2012, 42–45). In contrast to the anti-hegemonic English

School tradition,[3] Clark (2009a, 2009b) suggests that, by analogy to the role of great powers, hegemony is a potential institution of international society. From that perspective, a hegemon *can* have responsibilities; in the absence of other powers, the lone superpower can, at least in principle, define and exercise its responsibilities alone. In discussions about the global responsibilities of the hegemonic post-Cold War United States, Chris Brown (2004, 11–12) distinguishes unilateralists from multilateralists. Although both camps maintain that hegemonic status burdens the United States with great responsibilities, he observes, they differ significantly in their views on the nature of those responsibilities. On the one hand, multilateralists argue that any sole superpower is urged to cooperate with smaller states because it cannot resolve global problems singlehandedly and that the United States, for example, thus has a responsibility to direct the promotion of the public good (ibid., 11–12). On the other, unilateralists argue that the United States has to exploit its power to promote its own values and concepts of what is good to the world. For unilateralists, the outcomes of US policies have been critical, and stability and order have had no value per se (ibid., 12–13). The debate between unilateralists and multilateralists informs understandings of US expectations of China's global responsibility as China's influence blossoms on the international stage. Because permanent membership in the UN Security Council has been unable to entrench more profound consensus regarding the collective responsibilities of great powers, the ways in which single great powers or sole superpowers define their own responsibilities to other states, other (potential) great powers and even world society matter. It is therefore crucial to elucidate the ways in which China defines and interprets its emerging great power responsibility.

Pluralism, solidarism and great power climate responsibility

A basic tenet of the English School is that responsibility for managing international society rests largely on great powers. In general, both pluralists and solidarists agree that great powers have a special responsibility to ensure the achievement of international society's ultimate goals given their special role in that society. However, because pluralists and solidarists maintain divergent views on what those ultimate goals are, as well as how they should be promoted and accomplished, their stances regarding how and why great powers ought to shoulder their responsibilities also diverge. Given their focus on international order as a key value and means to promote the common good of international society, pluralists stress the functional responsibilities of great powers in that society, whereas solidarists, in underscoring the social attributes of power and responsibility, hold that great powers have a special responsibility to promote international justice and universal human values.

The pluralist camp of the English School emphasises that great powers have a special collective responsibility to 'ensure that the conditions of international peace and security are upheld' (Jackson 2000, 203). Slightly sceptical of the solidarist motives of great powers, Wight (1999 [1946], 42), for example, encourages observers to 'ask whose security is in question, and at whose expense it is purchased'. Since international order constitutes a key means to facilitate peaceful co-existence and international society's other ultimate goals, maintaining that order is the primary *functional* responsibility of great powers. On the one hand, such responsibility means that great powers need to pursue their interests prudently; they have to manage their relationship with one another and avoid harming other states and the functioning of international society (Bull 2002 [1977], 200; Watson 1982, 201). It therefore also means that great powers have to act in compliance with international law (Aslam 2013, 13). On the other, great power responsibility additionally means mediating international conflicts and preserving the general balance of international society. In suddenly intense conflicts, chief responsibility for peace negotiations falls to great powers, which have to 'agree at least tacitly on a form of crisis management' (Watson 1982, 201). When confrontation between great powers on opposing sides of a conflict is unavoidable, the powers themselves, not 'smaller and more immediate protagonists', are responsible for avoiding the use of force (ibid.). Of course, peace management can sometimes involve using force; great powers may, or even sense a moral duty to, use punitive measures to defend international peace and order if necessary. From a pluralist perspective, the US bombing of Hiroshima and Nagasaki in 1945, for example, can thus be interpreted as the materialisation of the US responsibility to promote the survival of international society, even if solidarists would argue that using nuclear weapons is anything but responsible behaviour. By the same token, China's reluctance to commit to engaging in coercive measures as part of the UN Security Council can also be viewed as an important reason for its incomplete acceptance as a responsible member of the great power club.

Because the solidarist camp of the English School espouses a so-called 'thicker' morality in international society, they demand that humans as well as states be viewed as members of that society. In other words, solidarists argue that individual humans around the world should be considered as moral referent objects of state responsibility, including great power responsibility. In practice, that view implies that great powers have great *diplomatic* responsibilities; they should advance international justice by using diplomatic tools to reach consensus about human values globally. Solidarists promote universal ideas such as human rights, the rule of law and good governance as 'new standards of civilization' that great powers ought to advance in their diplomacy. To some extent, that humanitarian responsibility of states has been recognised by contemporary international law (Knudsen 2016), in which a pivotal development was the adoption of the principle of the responsibility to protect during the 2000s. According to that principle, if rogue states violate the human rights of their citizens, great powers are expected to bear the greatest responsibility to interfere. Nevertheless, the extent to which certain universal values exist remains an important question. Normative theories of international relations, including that of the English's School's solidarist camp, tend to be very Eurocentric, which has sparked criticism among non-Western scholars and practitioners (Hurrell 2016). In discussions regarding China's rise, it indeed seems that the 'features of the New Standard of Civilization' (Fidler 2001, 150) form a normative basis for Western criticism of contemporary China. In short, because China is not a democratic country, it is not part of 'us' but other (Zhang 2011) and thus cannot be fully accepted in the great power club. To cope with the rise of China and other non-Western emerging powers, international society may need to adjust its rules and conceptualisations of justice, which could also gradually alter ideas about international rights and responsibilities.

Compared to the Eurocentric bias of solidarism, state-centric solidarism takes a rather culturally sensitive approach to international ethics. It maintains that the ultimate goal of international society is to promote human wellbeing but does not set preconditions for how, and by whom, such wellbeing should be defined. On the contrary, it stresses the importance of processes of responsibilisation in social life – that what *human wellbeing* means in practice in different settings is specific to context. Via processes of responsibilisation, participants of social practices, including great powers, negotiate what value they place upon certain aspects of certain actors' wellbeing. In international politics, *development* has been the key term in shaping processes of responsibilisation since the inauguration speech of US president Harry Truman in 1949. Truman's speech was based on an idea that 'all the peoples of the world were moving along the same track, some faster, some slower, but all in the same direction' (Sachs 1993, 4). Truman characterised Western models of socioeconomic development as universal norms and resonated with Western beliefs of progress and improvement that the future will, or at least should, be better than the present (Barry 1999). However, such ideas of progress did not specify for whom and in what terms the future would be better: for (Western) political elites, all humans or all living creatures? Truman's conceptualisation also lumped diverse African, Asian and Latin American countries into 'one single category – the underdeveloped' (Sachs 1993, 4). Despite those criticisms, human wellbeing largely continues to be measured in economic terms, including gross domestic product, and states tend to treat economic growth as their principal responsibility. By extension, international practices focusing on economic growth have also dictated international climate practices. The negative impacts of climate change are often discussed in economic terms, whereas other, qualitative aspects of human wellbeing are dismissed. Because real human suffering caused by climate change cannot be estimated in economic costs, however, Roberts and Parks (2007) propose more appropriate measures of climate-related risks, including the number of people killed, made homeless or otherwise harmed by climate-related disasters such as floods, heatwaves, droughts and windstorms in individual states.

Within the English School, surprisingly little attention has been paid to climate ethics. Arguably, the pluralist perspective on climate change would focus on state-centric harms and risks to international order. Several studies and reports have indeed highlighted that climate change poses national security risks in every country worldwide (e.g. Mazo 2010). For some states, such as small island nations in the Asia–Pacific, climate change even poses an existential threat, for when sea levels rise due to the melting of ice in polar areas, they will literally disappear from the world map – a phenomenon that Milla Vaha (2015) calls *state-extinction*. Given growing consensus that climate change is a potential source of international conflict, it is reasonable to assume that climate change risks the maintenance of international order. Moreover, given their managerial role in international society, great powers can be assumed to have a functional responsibility to lead efforts to mitigate climate change as a means to maintain international peace and security. By some contrast, solidarists would add human suffering to the list of harms caused by the adverse effects of climate change. Climate change is indeed inherently an issue of international justice, for especially vulnerable to climate change are the poorest people living in developing countries that lack sufficient resources to adapt to such change and, often located in tropical and sub-tropical areas, are most likely to be affected by it.

For many scholars in the English School, international law ranks among the most important primary institutions of international society. Because international law captures the shared rules of co-existence accepted by members of international society, it plays a critical role in the construction of that society. By specifying what constitutes acceptable and unacceptable conduct therein, international law enables and constrains the actions of states, including great powers. Upholding international law is an important attribute of great power responsibility; great powers must follow its principles in order to maintain international order and legalise their hegemonic status (Simpson 2004). However, maintaining international order sometimes requires the violation of international law. As Bull (2002 [1977], 138) points out, the United Kingdom and France did not criticise Russia for attacking Finland in 1939, despite the action's clear violation of international law, because the attack stabilised the European balance of power. According to pluralist ethics, great powers thus have a responsibility to act against international law if necessary to maintain international order. By comparison, solidarists observe that, as all laws, international law is a human construction and thus reflects power relations. International law can therefore be imperfect and unfair and undermine fundamental human values. If the justice of legal rules were challenged for humanitarian reasons, for example, solidarists would consider a great power's illegal actions to be legitimate. However, because an illegal action by a great power can be legitimate only if based on a consensus in international society that the action is indeed necessary, 'consensus is the benchmark of legitimacy' (Clark 2005, 164). Since great powers tend to play an influential role in law-making processes in international society and incorporate special rights into those processes, they are usually content with and eager to enforce them (Simpson 2004, 70). Among other incentives for great powers to commit to international law, they typically prefer to maintain their hegemonic role by means of law instead of force (Onuma 2003, 117).

Although the violation of international law does not warrant legal sanctions for great powers, it is incorrect to assume that great powers would readily commit such violations. Breaking international rules can seriously damage the image of a great power and, in turn, hinder its pursuit of national interests and jeopardise its international leadership role in the long term. Moreover, at least in liberal-democratic great power states, domestic pressure to respect international norms and rules is quite high (Onuma 2003, 119). Even though US president Donald Trump clarified in June 2017 that his administration would not commit to international climate norms, he did not simply abandon the United Nations Framework Convention on Climate Change but chose to implement the withdrawal in accordance to the Paris Agreement itself. In effect, the United States will be part of the accord until 2020 and continue to implement national climate mitigation plans compiled by the Obama administration. Accordingly, the United States sent a small delegation to the UN climate negotiations in Bonn in November 2017. If the Trump administration at least attempts to meet US domestic climate targets and reports its actions, it will not violate international law. In that case, pluralists would maintain that the Trump administration will have fulfilled its functional great power responsibility to pursue the common good of international society by ensuring that it has met the rules of international law and not jeopardised international order. At the same time, it remains questionable whether Trump's decision to dismiss efforts to allay climate change increases the risk of international conflict and harms humanitarian security around the world. Given strong evidence of the relationship between climate change and conflicts, some pluralists would likely advocate the functional responsibility of great powers to mitigate climate change and thus criticise Trump's decision. Solidarists, by contrast, would undoubtedly condemn Trump's decision as irresponsible because it dismisses the common good of humans worldwide.

International law articulates only a minimum standard of conduct, and responsible states do more. In particular, solidarists expect responsible great powers to willingly uphold more than international law by promoting human wellbeing globally. International law is thus not the only metric of international responsibility. To some extent, the Obama administration fulfilled its diplomatic responsibility to promote human values by supporting the mitigation of climate change; although the United States had not ratified the Kyoto Protocol, Obama's climate diplomacy worked to achieve consensus on climate responsibility and the adoption of the Paris Agreement. By contrast, Trump's decision to withdraw the United States from the Paris Agreement has faced wide international criticism although the decision did not violate international law, and the global leadership of the United States has been increasingly questioned as a result. Trump's climate scepticism has also generated widespread discontent in the United States, and in response, many subnational and civil society actors have become more active advocators of climate responsibility than ever before. Indeed, the coalition of non-state actors that issued America's Pledge in 2017 represented more than half of the US economy and even argued that *they* should represent the United States in international climate negotiations (Neslen 2017). From the perspective of the English School, that proposition raises the question of whether the expanding role of non-state actors in international society can induce structural changes therein.

To reiterate, legitimacy is an important aspect of great power responsibility for both pluralists and solidarists. As Ian Clark (2005) points out, principles of legitimacy determine which actors are considered to have the right to participate in international practices and what is viewed as appropriate standards of conduct in international society. The English School has thus tended to underscore the importance of a great power's legitimate *international* conduct when assessing its rightful membership in the great power club. By contrast, Gareth Evans's conceptualisation of responsible international citizenship, introduced in a serious of foreign policy speeches the late 1980s and early 1990s, maintains that states cannot fulfil their international responsibilities if they do not fulfil those responsibilities at home as well. That dynamic becomes especially clear in the context of international climate politics, for although a state may play a constructive role in international climate negotiations by endorsing international cooperation and committing to international agreements, it cannot be deemed a responsible international actor if it does not meet its responsibilities by taking *domestic* actions. At the same time, even if not a signatory to an international climate agreement, a state may nevertheless prove to be a responsible international actor by undertaking ambitious domestic measures to mitigate climate change. In that case, addressing climate change presumably aligns with a state's national interests, which is an adequate justification for pluralist ethics. Indeed, politicians often highlight the domestic economic benefits of their climate policies. President Obama, for example, despite having declared climate change a top priority for his administration, did not emphasise the international responsibilities of the United States when addressing US audiences (Bukovansky et al. 2012, 153). From a solidarist perspective, however, economic interests are not an appropriate or legitimate basis for great power responsibility, which has to be bound to human values and international justice.

Clearly, both domestic and international legitimacy are important factors of great power responsibility. Whereas Bull and other pluralists would highlight the functional importance of the legitimate conduct of responsible international citizens both at home and abroad, solidarists would call for more robust definitions of *responsible international citizenship* that stress attention to the social elements of legitimacy at the domestic and international levels. That observation can be articulated in Weberian terms as well. Max Weber conceives a 'co-constitutive relationship between the domestic and international realms' (Hobson & Seabrooke 2001, 240), and his approach captures the state-centric solidarist conception of responsibility; the 'domestic norms of impartiality and fairness entwined with the ethic of responsibility, coupled with the domestic – and international – social balance of power' all unite in his theory of the state in particular and of international society in general (ibid., 269). Given Weber's sophisticated analysis of state–society relations emphasising a strong civil society, however, his approach does not may not fully explain China's membership in the great power club.

Practices of great power responsibility

For as long as it has existed, international society has doubtlessly always had great powers. In the terms of the English School, however, such powers have constituted an international club of 'legalized hegemony' only since the early nineteenth century (Reus-Smit 1999, 109; Simpson 2004, 73). At the Congress of Vienna (1814/15), Austria, Great Britain, France, Prussia and Russia – the four major powers that had defeated Napoleon – plus the restored King of France each obtained the recognised status of great power and established the Holy Alliance as an ideological basis for the rule of their great power club (Simpson 2004, 96–115; Brown 2004, 7). To quote Brown (2004, 7), the five great powers were 'conscious of themselves as constituting an institution which was separate from other states and in possession of special responsibilities as well as rights vis-à-vis international society'. According to F. R. Bridge and R. Bullen, such status formed an implicit social contract between great powers and smaller states; 'just as the great powers claimed special rights for themselves, so the small states claimed that the great had special responsibilities for their well-being' (quoted in Bukovansky et al. 2012, 27). Because the contract was not written in international law, however, great power responsibility was an informal norm at the time. Later, the League of Nations also granted special status to great powers but did not form an 'institutional/ideological unity' similar to the Concert of Europe (Brown 2004, 8).[4] Not until 1933, when Great Britain, France, Germany and Italy signed the Four Power Pact that declared them to be 'conscious of the[ir] special responsibilities' (Bukovansky et al. 2012, 29), was great power responsibility formalised.

In the 1940s, great power responsibility experienced several influential developments. In 1943, then British foreign secretary Anthony Eden declared that 'special responsibilities do rest on our three powers', meaning the United Kingdom, the Soviet Union and the United States (quoted in Bukovansky et al. 2012, 29). The following year, he called for the formalisation of the special responsibilities of great powers by establishing a new world organisation that would 'make it possible for them [the Four Powers] to carry out the responsibilities which they will have agreed to undertake' and that 'they must be given … a special position in the organisation' in the process (quoted in Bukovansky et al. 2012, 30). As a result, 'everyone' was talking about responsibility by 1945 (cf. Bukovansky et al. 2012, 29–30). For example, a night before his death, President Franklin D. Roosevelt (1945) responded to the events of World War II by stating, 'Today, we have learned in the agony of war that great power involves great responsibility'. Similarly, his successor, President Truman told the US Congress – and reiterated it at the UN General Assembly in San Francisco in April 1945 – that 'While these great states have a special responsibility to enforce the peace … . The responsibility of the great states is to serve, and not dominate the peoples of the world' (Truman 1945).

In its contemporary form, the great power club was institutionalised by the establishment of the UN Security Council in 1945. Under the UN Charter (1945, ch. 5, Article 24), great powers were formally appointed to have special responsibilities, including 'primary responsibility for the maintenance of international peace and security'. Interestingly, the formalisation of such responsibility was not actively sought by great powers, as their seemingly genuine belief that international society had imposed a burden upon them suggested (Simpson 2004, 170). At the time, great power responsibility was largely based on great powers' material capabilities because, according to Eden, 'the more power and responsibility can be made corresponding, the more likely it is that the machinery will be able to fulfil its functions' (quoted in Bukovansky et al. 2012, 31). The permanent membership of the United States, the Soviet Union or Russia, Great Britain, France and the Republic of China (P5) at the UN Security Council – and especially their veto rights – made them 'morally superior' to international doctrines such as equality and unanimity and placed them 'above the law they are to impose on others' (Wight 1999 [1946], 45). Such privilege made the great powers special, which, in turn, was viewed as the source of their responsibility (Brown 2004, 9).

When the Cold War broke out soon after the establishment of the UN, the institutionalisation of the P5 did not realise a collective understanding of the responsibilities of great powers in practice. During the Cold War, being in the P5 carried symbolic status only and indicated neither power nor responsibility, as illustrated by the Republic of China's (i.e. Taiwan's) maintenance of its status as a permanent member of the UN Security Council for 20 years after its regime's collapse in mainland China (Brown 2004, 9). In Western international society, the United States assumed a 'new position of world responsibility' and became the 'principal protector of the free world' (Truman 1948). That responsibility, however, was not bound to the collective responsibility of the P5 but the global capabilities of the United States as a singular hegemonic power (Bukovansky et al. 2012, 34; Brown 2004, 11–13; Clark 2011, ch. 6; Ikenberry 2009, 76–79). Later, the dismantling of the Soviet Union and the end of the Cold War afforded new possibilities to the P5 to fulfil their responsibilities in promoting international peace and security originally articulated in 1945. Many new concepts, including Gareth Evans's idea of good international citizenship and Francis M. Deng and his colleagues' notion of sovereignty as responsibility, gave rise to wider debate over ethics and foreign policy in the late 1980s and early 1990s. Consequently, human rights emerged as a new 'standard of civilization' (Donnelly 1998) that conceptualised the responsibility to protect and the willingness to undertake humanitarian intervention as key attributes of great power responsibility (e.g. Wheeler 2000).

Ideas about environmental security as an approach to international security first surfaced in the early post-Cold War era.[5] Although scholars initially focused on environmentally induced conflicts, by the end of the Cold War the UN Security Council began to pursue a more expansive approach to international security. In 1992, it noted that 'non-military sources of instability in the economic, social, humanitarian and ecological fields have become threats to peace and security' (UN Security Council 1992). Since the mid-2000s, when many 'securitizing moves' to promote climate change mitigation were made (Trombetta 2008, 594–595), the relationship between climate change and violent conflict has been widely studied (e.g. Lee 2009; Mazo 2010; Welzer 2012). In effect, those developments have generated debate about the UN Security Council's role in mitigating climate change, which, if viewed as a threat to international peace and security, arguably can and *should* be added to the organisation's agenda.

In 2007, with the British presiding, the UN Security Council held the first-ever debate on the relationships among climate change, energy and security, although some members, including China, doubted whether the occasion was the appropriate forum for the discussion (United Nations 2007). Nevertheless, British foreign secretary and president of the Council Margaret Beckett insisted that the members discuss the security-related impacts of climate change, because the 'Council's responsibility was [is] the maintenance of international peace and security, and climate change [has] exacerbated many threats, including conflict and access to energy and food' (ibid.). The UN General Assembly (2009a) encouraged relevant UN organisations to intensify their efforts to allay climate change, 'including its possible security implications', and asked the UN Secretary-General to submit a comprehensive report addressing the potential security-related impacts of climate change. In response to the UN General Assembly, the UN Secretary-General's report defined *climate change* as a threat multiplier that could affect security by increasing vulnerability, hindering development, necessitating increased coping and security, promoting statelessness and engendering international conflict (UN General Assembly 2009b). In 2011, the UN Security Council, with Germany presiding, also discussed the potential security-related impacts of climate change and consequently adopted its first-ever statement on the issue (UN Security Council 2011). However, the body made no decision regarding whether new environmental peacekeeping forces, so-called 'green helmets', could be used to manage conflicts caused by resource scarcity (United Nations 2011). In 2013, the UN Security Council held informal talks addressing the issue but failed to define *climate change* as an international security threat due to resistance from China and Russia (Krause-Jackson 2013). The following year, however, President Obama (2014) explicitly acknowledged the link between great power responsibility and climate change. Altogether, though the UN Security Council has not made any concrete decisions about climate change, that it has discussed climate security has upgraded the status of climate change on the global political agenda. As an environmental issue, climate change is now a matter of soft politics, while its potential securitisation makes it a part of hard politics as well. That development could signal that climate responsibility is emerging an attribute of great power responsibility.

Expectations of China's responsibility

For the time being, China has no salient identity as a great power. On the one hand, China's increasing wealth generates expectations of greater international respect, and it no longer accepts being left on the periphery of international society but it struggles to be recognised as a great power. On the other, its status as a developing country continues to be central to its identity, especially in international climate politics. Similarly to individuals, states construct their identities in social interaction by engaging in various international practices; when newcomers join, as China gradually has into the great power club, they learn new ideas and ways of operating in that world, which consequently transforms their identity.[6] Other participants in a practice who have a 'stake in making up certain social categories and in trying to make people [states] conform to them' (Zalewski & Enloe 1995, 282) play an important role in shaping state identity. In general, both China and the United States agree that world peace is an essential value of international society and that great powers have a responsibility to maintain global peace. However, the two states seemingly have different views on the other global responsibilities of great powers. When a rising China asks itself 'Who am I?'[7], the United States tries to influence the answer by (re)defining what it means to be a great power in the twenty-first century and what sorts of responsibilities accompany that status. Although not all US contributions have explicitly defined *responsibility* as a rule of membership in the great power club, there is a clear tendency, as I later demonstrate, that responsibility is a central requirement for states that seek recognition as great powers. As Buzan (2004, 67) points out, the 'key here is not just what states say about themselves and others' but 'how they behave in a wider sense, and how that behaviour is treated by others'. As a result of that learning process, 'relative newcomers become relative old-timers' over the course of time (Wenger 1998, 90). Such advancement is usually unmarked and implicit; suddenly, a state finds that it has risen to a position at which it can educate new newcomers, and other participants have begun to expect the state to know and do more than it is already certain it does. That development is exactly what seems to have happened amid China's rise during recent decades; the West has expected China to shoulder greater global responsibilities, whereas China continues to regard itself as a developing country unable to respond to those new demands (Kopra 2016). Before exploring the discursive clash between the United States and China over China's global responsibility, I briefly outline China's entrance into the great power club.

At the founding of the UN Security Council, the Republic of China was given a permanent seat, although not without dispute about whether it merited such a status (Simpson 2004, 173). The position did not result in China's self-perception as a great power because the People's Republic of China (PRC) had neither international rights nor legitimate representation at the UN until October 1971. In 1972, US president Richard Nixon's visit to the PRC re-established US relations with China, largely based on their 'mutual antipathy towards Moscow' (Lanteigne 2013, 105). Initially, US leadership was optimistic about China's reforms and assumed that 'China would learn to be more like us'. However, the Tiananmen Square incident in 1989 profoundly altered the US policy towards China, and a 'containment policy' was applied until 1993, when it was replaced with an 'engagement policy' (Zheng 1999, 126). After the Taiwan Strait crisis (1995–1996), the Clinton administration announced that its long-term objective for China was to integrate the country into international society 'with all the privileges and responsibilities of a major power' (ibid., 128). In 1995, then US secretary of defense William Perry noted that the engagement strategy would ensure that China would become a responsible member of that society (Jin 2011, 11). In effect, the strategy meant that the United States would commit to helping China to join the great power club and that China would respect established international rules and act accordingly. Despite Chinese scepticism over US motivations concerning China, the engagement policy afforded the Chinese state a way to emerge as a real great power (Zheng 1999, 128–129). Chinese president Jiang Zemin's visit to Washington, DC, in 1997 and US president Bill Clinton's visit to Beijing in 1998 restored the official dialogue between the countries' leaders that had been abandoned since the Tiananmen Square incident in 1989 (Harding 1999, 7–8). In that dialogue, the leaders decided to strive for a 'constructive strategic partnership', which they characterised as a 'goal to be pursued, not an accomplishment that could be celebrated' (ibid., 21). The two summits marked the end of a 'decade of flux in great power relations' and catalysed the development of China's great power identity (Rozman 1999, 383).

During his presidential campaign and early in his presidency, US president George W. Bush took a more hard-line policy towards China and redefined what was once a US–Chinese partnership as a competition with a power that should be 'treated without ill will but without illusions' at the same time (Federation of American Scientists 1999; cf. Yu 2009, 84). Nevertheless, Bush supported China's accession to the World Trade Organization and its bid to host the 2008 Olympic Games in Beijing, which relieved tensions after a naval aircraft collision between the militaries of the two countries in the South China Sea in April 2001. Immediately after the incident was resolved, Bush committed to establishing 'constructive relations' with China (Yu 2009, 87), and in return, China supported Bush's post-September 11 'war on terror'.

Although the mutual cooperation improved China's relations with the United States, it did not award China the status of strategic partner with the United States (Roy 2002; Shambaugh 2002). Bush, however, seemed determined to further improve US–Chinese relations and promote multilateralism. For instance, he attended the Asia-Pacific Economic Cooperation in Shanghai in October 2001, despite debate over the importance of the meeting when the United States was at war. Bush also celebrated the thirtieth anniversary of President Nixon's visit to China by visiting China again and took a 'symbolic step forward in the place where Nixon stopped' while touring the Great Wall (Yu 2009, 87–88). A more concrete step in US–Chinese relations was taken when China's government approved applications of the Federal Bureau of Investigation (FBI) and Central Intelligence Agency (CIA) to enter China and begin operating in Beijing in 2002 (Zhang & Zheng 2012, 628).

In 2005, then US deputy secretary of state Robert B. Zoellick introduced the concept of responsible stakeholder to international politics. According to Zoellick (2005), 'All nations conduct diplomacy to promote their national interests. Responsible stakeholders go further: They recognize that the international system sustains their peaceful prosperity, so they work to sustain that system.' Although Zoellick's speech did not clearly define *responsible stakeholder*, it nevertheless spurred international debate over expectations of China's global responsibilities. Arguably, the primary goal of introducing the concept was to describe China's international responsibilities in the context of US interests and expectations, as well as to urge China to fulfil those responsibilities (Gill 2007). In any case, Zoellick's (2005) concept was based on pluralist ethics, for it stressed that, as a member of international society, China has a responsibility to strengthen the international system that has made its rise possible. At the same time, the concept suggested, China should not challenge the existing rules of international society or promote competing norms or another type of international order. In general, Zoellick was optimistic about China's potential to become a responsible stakeholder and encouraged the United States to cooperate with China in that regard.

The following year, the concept of responsible stakeholder was written into the US National Security Strategy 2006, which demanded that 'As China becomes a global player, it must act as a responsible stakeholder that fulfils its obligations and works with the United States and others to advance the international system that has enabled its success' (White House 2006). The first Obama administration adopted similar views, and Zoellick's successor as US deputy secretary of state, James Steinberg, formulated his own China paradigm, dubbed 'strategic reassurance', in 2009. Steinberg (2009) emphasised China's negative responsibility not to harm other states:

> Just as … we are prepared to welcome China's 'arrival' … as a prosperous and successful power, China must reassure the rest of the world that its development and growing global role will not come at the expense of security and well-being of others.

Steinberg (2009) urged China to reassure other states that it poses no international threat: 'When it comes to the international system, we must ensure that new powers like China – and there are others as well, of course – can take their rightful place at the table without generating fear or mistrust'. Although he confirmed that the United States was 'ready to accept a growing role for China on the international stage', he stressed that it 'will also be looking for signs and signals of reassurance from China' and if 'China is going to take its rightful place, it must make those signals clear' (ibid.). In contrast to Zoellick, who made no reference to climate change or environmental issues in his speech,[8] Steinberg (2009) mentioned that US–Chinese cooperation towards mitigating climate change was necessary due to their statuses as the world's largest carbon emitters. The Obama administration also underscored that China's increasing capacity should be accompanied with broader positive responsibilities. For example, President Barack Obama welcomed China's greater global role, 'in which a growing economy is joined by growing responsibilities' (White House 2009). After China became the world's largest carbon dioxide emitter in 2006, the United States began to urge China to shoulder more responsibility in curbing climate change as well.

During the second Obama administration, constructing a 'productive and constructive relationship' with China was a major strategic goal for the United States. In a speech delivered a week after President Obama's re-election in November 2012, then US national security adviser Thomas Donilon (2012) urged 'Beijing to define its national interest more in terms of common global concerns and to take responsibility for helping the international community address global problems'. He also asked China to become a responsible international citizen: 'Now, we've been clear that as China takes a seat at a growing number of international tables, it needs to assume responsibilities commensurate with its growing global economic impact and its national capabilities' (ibid.). In March 2013, Donilon reiterated that demand and encouraged US–Chinese cooperation 'to build a new model of relations between an existing power and an emerging one'. He stressed that no natural law exists that could determine that 'a rising power and an established power are somehow destined for conflict' (Donilon 2013).

In November 2016, when Donald J. Trump was elected as US president, the US began to play a starkly different role in international society and in its relations with China. It had become clear during his presidential campaign that Trump would pay less attention to great power responsibility and focus on national politics instead. In particular, he had called climate change 'a Chinese hoax' on Twitter in 2012, which fuelled international fears for the continuance of the US leadership in international climate politics. Although I discuss Trump's hostile attitude towards climate politics in greater detail later, here it is necessary to emphasise how Trump's unwillingness to shoulder climate responsibility and sign an international trade agreement called the Trans-Pacific Partnership consequently elevated China's global status. In the light of Trump's irresponsible international policy actions, China has begun to be viewed by the world as a more responsible international player than ever before. Although Trump's election thus altered the international expectations of China's great power responsibility, it remains unclear whether or not China is willing or able to live up to those expectations and shoulder more responsibility in international politics. Chinese leaders, for example, have already stated that the West needs to stop invoking 'China responsibility theories' that exaggerate the country's duty to diffuse the nuclear threat on the Korean Peninsula (Reuters 2017). Regarding international climate politics, the world could receive a similar message, one backed with practical evidence, for China's national circumstances have not changed to suddenly afford it more resources, know-how or political willingness to lead the charge against climate change. An important factor in those policy decisions is how China constructs its identity as an emerging great power and perceives its corresponding great power responsibility.

China's emerging notions of great power responsibility

In the years following the PRC's establishment in 1949, the Chinese government was keener to develop alternative international practices than join the great power club (cf. Foot 2001, 24–28). To join the UN in 1971 and the Bretton Woods institutions in 1980, however, the PRC had to normalise its relations with the United States. Early during China's reform era in the late 1970s, the Chinese admired the United States as a 'symbol of a comfortable material life' and for its 'rational institutional arrangements, and advanced technologies' (Zheng, 1999, 51–52). During the 1980s, however, when the Chinese discovered that the West was 'far from their original high expectations' and that its practices towards China were unfair, a new sort of nationalism began to emerge in China (ibid.). Evan S. Medeiros (2009, 95) claims that, since the early 1990s, three tenets have guided China's relations with major powers – 'non-alliance, non-confrontation, and not directed against any third party' (不结盟, 不对抗, 不针对第三方) – all naturally undergirded by economic interests. In accordance with those tenets, China has not formed alliances but partnerships around the world since the end of Cold War.[9] At present, China maintains a strategic partnership with the European Union, Russia and the United States. Efforts to constrain US global influence as well as Japan's regional influence, however, remain central to China's foreign policy. In particular, contemporary Chinese leaders have promoted an Asia for Asians policy and introduced new concepts such as the Asia-Pacific Dream (亚太梦) that can be conceived as countermoves to the so-called 'Asia Pivot' of the United States. Articulated by Deng Xiaoping in the early 1990s, China's strategic guidelines for US–Chinese relations continue to be to 'increase trust, reduce problems, strengthen cooperation, and avoid confrontation' (增加信任, 减少麻烦, 加强合作, 不搞对 抗) (Medeiros 2009, 98).

Since the late 1990s, Chinese intellectuals have debated at length the international role and expectations of China.[10] In his speech to the Russian State

Duma in 1997, Chinese president Jiang Zemin acknowledged that great powers have great responsibilities by declaring that 'being major powers of influence and permanent members of the UN Security Council, China and Russia shoulder an important responsibility for safeguarding world peace and stability' (quoted in Yeophantong 2013, 331). Since then, Chinese intellectuals have proposed that China, as a nuclear power and permanent member of the UN Security Council, should redefine its national interests to meet international expectations regarding its responsibility (Yeophantong 2013, 348). Moreover, following his analysis of internal and external factors influencing whether China will become a 'responsible great power' in the twenty-first century,[11] Xia Liping (2001, 17) identified 'some conditions necessary to make China a responsible great power', including that China 'should: (1) play its role in international society not only according to its national interests, but also in order to benefit regional and world peace, development, stability, and prosperity; (2) take its international obligations more seriously; and (3) participate in the formulation of international rules'.

During his visit to the United States in 2002, Zheng Bijian, former executive vice-president of the Central Committee's Central Party School, observed that Americans had severe doubts about China's rise to great power status, something which would later impede Sino–American relations and China's pursuit of great power status (Glaser & Medeiros 2007, 294). Consequently, in 2003 Zheng introduced the concept of 'peaceful rise' (和平崛起) to dispel fears about Chinese threat, and the following year, the concept was adopted by the Hu-Wen administration as a new national strategy. However, the 'rise' part of the concept was quickly judged to be counterproductive and soon replaced with 'development'.[12] Since 2004, the concept of 'peaceful development' (和平发展), as the leading principle of Chinese foreign policy, has assured the world that China's rise will be peaceful and that no 'hegemonic war' will occur. In general, both China's government and Chinese scholars embraced Zoellick's conception of responsible stakeholder, although some factions pondered whether it was an engagement policy or a containment policy in nature (Jin 2006; Masuda 2009, 67). No official remarks on the concept were issued, and no explicit commentary appeared in the *People's Daily,* the mouthpiece of the CCP (Masuda 2009, 67). At his meeting with President Bush in 2006, President Hu commented that 'China and the United States are not only stakeholders, but they should also be constructive partners' (quoted in Yu 2009, 97). On the one hand, Hu seemed to accept the characterisation of China as an 'international stakeholder' because it promoted China's international status. On the other, he did not relate it to responsibility, likely given his incomplete approval of the US understanding of China's global role (Masuda 2009, 67).

Shortly after Zoellick's speech, the State Council Information Office (2005) issued a white paper titled 'China's Peaceful Development Road' to elaborate upon the country's philosophy of peaceful development. The paper highlighted China's development-related needs and declared that its 'development will never pose a threat to anyone' because 'peaceful development is the inevitable way for China's modernization'. It also assured that, 'Active in the settlement of serious international and regional problems, China shoulders broad international obligations, and plays a responsible and constructive role'. Although the paper suggested that 'China is certain to make more contributions to the lofty cause of peace and the development of mankind', it shifted primary global responsibilities to developed countries, which, it stated, 'should shoulder greater responsibility for a universal, coordinated and balanced development of the world', whereas 'developing countries should make full use of their own advantages to achieve development'. China's second white paper on peaceful development reminded the world that China is 'actively living up to international responsibility' (Information Office of the State Council of the People's Republic of China 2011), underscored China's status as a developing country and suggested that China should not be expected to shoulder greater global responsibilities until it meets domestic challenges and achieves a higher level of development. However, the second paper did not indicate what level of development China should achieve before it assumes more global responsibility, nor when China's government expected that level to be reached.

As Jin Canrong (2011, 12) describes it, being a great power means setting international agendas proactively and not allowing other states to control agendas or define global responsibilities. Accordingly, the Chinese Communist Party (CCP) has begun to develop and promote its own concepts, including the 'harmonious world' (和谐世界), the 'China dream' (中国梦), the 'Asia-Pacific dream' (亚太梦), a 'new type of major country relationship' (新型大国关系) and a 'new type of international relations' (新型国际关系), as means to organise international society. Of course, only time will tell whether those concepts can reorganise international practices to become less Westernised and accommodate Chinese values and interests more efficiently (Kopra 2016, 30). The purpose of the concepts seems to be to reform international society in a 'responsible manner', not to replace existing practices from which China has benefitted (cf. Buzan 2010, 29–33). In international finance, China has introduced sources of global governance by establishing new multilateral mechanisms such as the Asian Infrastructure Investment Bank and the New Development Bank, which are arguably alternatives to the International Monetary Fund and the World Bank. Moreover, President Xi Jinping's signature New Silk Road initiative 'One Belt, One Road' (一带一路) could have far-reaching political impacts in the coming years.

From this book's perspective, the concept of the 'new type of great power relationship' first expressed by China's then vice president Xi Jinping in February 2012 is critical. Xi (2012) claimed that

> We [the United States and China] should work hard to implement the agreement between the two presidents, expand our shared interests and mutually beneficial cooperation, strive for new progress in building our

> cooperative partnership and make it a new type of relationship between major countries in the 21st century.

Xi identified four ways in which two countries should collaborate in order to foster the described new type of relationship: increasing 'mutual understanding and strategic trust', respecting 'each side's core interests and major concerns', deepening 'mutually beneficial cooperation' and enhancing 'cooperation and coordination in international affairs and on global issues', including climate change. Moreover, Xi (2012) declared:

> Our world is undergoing complex and profound changes. China and the United States should meet challenges together and share responsibilities in international affairs. This is what China–US cooperative partnership calls for and what the international community expects from us.

A couple of months later, then president Hu Jintao (2012a) reiterated the call for a 'new type of great power relationship' and emphasised the importance of mutual trust. He stated that the 'world we live in is big enough for China, the United States and all other countries to achieve common development'. In his report to the 18th National Congress of the CCP in 2012, Hu (2012b) asserted that China would continue to 'play its due role of a major responsible country', and the new type of relationship was included as a goal in the 18th Party Congress work report. After his nomination as China's premier in 2013, Li Keqiang confirmed that the fifth generation of Chinese leadership would 'work with the Obama administration to work together to build a new type of relationship between great countries' (Jones & Lim 2013). That commitment indeed emerged as a key element of the Xi–Li administration.

In 2013, Chinese foreign minister Wang Yi (2013) gave a rare comprehensive statement of China's foreign policy titled 'Exploring the Path of Major-Country Diplomacy with Chinese Characteristics'. The following year, President Xi introduced the concept of 'major-country diplomacy with Chinese characteristics' at a high-level international conference in Beijing (Xinhua 2014). Although the official translation was thus 'major country diplomacy', the Chinese concept 大国外交 could be translated as 'great power diplomacy'. Wang pledged that China's fifth generation of leadership would take a more proactive approach to diplomacy. According to Wang (2013), China was 'ready to respond to this expectation of the international community … to undertake its due responsibilities and make greater contribution to world peace and common development'. He added:

> As a permanent member of the UN Security Council, China is always conscious of its international responsibilities and obligations and stands ready to offer more public goods and play its unique and positive role in addressing various issues and challenges in the world.
>
> (ibid.)

After Zoellick's speech in 2005, China has taken a more active part in UN peacekeeping operations, which can be viewed to signal its increasing acceptance to shoulder great power responsibility (Foot 2001; Suzuki 2008). Nevertheless, China has not fully accepted the Western concept of human rights and other attributes of great power responsibility from the US perspective (Kopra 2018). Therefore, as Pang Zhongying (2006, 9) notes, Wang's statement does not necessarily mean that China is 'fully prepared to embrace the notion that it is a custodian of the current international system, with all of the responsibilities that would entail'. In late 2014, Chinese Vice Premier Wang Yang confirmed that assumption: 'China and the US are global economic partners, but the leader of the world is the United States. The United States leads the system and rules; China is willing to join the system and to respect the rules and hopes to play a constructive role' (Chinaiiss 2014).

However, President Trump's election in November 2016 dramatically changed that characterisation. Trump had already clarified during his presidential campaign that his administration would no longer lead international society or follow its rules, as China and other states had expected it would. For China, that shift opened up a new opportunity to define and demonstrate how it perceives great power responsibility. For instance, President Xi seized an opportunity at the World Economic Forum in Davos in January 2017 to praise positive aspects of globalisation and portray China as the champion of free trade (World Economic Forum 2017). By October 2017, China had become even more confident about its new global position. In particular, in his speech to the 19th CCP Congress, Xi (2017) declared that China had entered a new era of power in which it leads the world on political, economic, military and environmental issues. Given his speech's extraordinarily strong emphasis on the development of military capabilities, it seems that Chinese leadership in the heralded new era is based upon not only economic power but also more traditional hard power. Xi also reminded the world that making 'new and greater contributions for mankind is our Party's abiding mission' but did not comment upon how China would pursue that mission in practice.

From the perspective of great power responsibility, neither President Xi's plans for China's new era nor his ideas about a new type of great power relationship have marked any breakthroughs. The new type of great power relationship focuses more or less on core interests, not common ones that could be translated into new responsibilities for China and the United States. Implicitly, the conceptualisation calls for hard power and an attempt to persuade the United States to respect China's sphere of interest in East Asia (Kopra 2016; Kopra 2018). In that light, international climate politics provides an interesting case in China's emerging notions of great power responsibility. Chinese leaders often refer to the massive size of the state when discussing its global responsibilities, and though they dub China a 'responsible big country' (负责任大国), the Chinese conception could also be translated as 'responsible great power'. According to Xi (2015), for example, 'Being a big country means shouldering greater responsibilities for regional and world peace and development, as opposed to seeking greater monopoly over regional and world affairs.' Moreover, it seems that China increasingly identifies itself as a great power with great responsibilities in international climate politics and has planned policy measures to meet those responsibilities (cf. Kopra 2016; Kopra 2018). Due to its 'international responsibilities and obligations as a new type of major country', China has not only issued all of the important climate policies in joint statements with the United States but also promised to give more financial and technological support to developing countries to assist them in fulfilling their climate objectives (China Daily 2016). At the same time, improving the country's status and developing the first principle continue to be important elements of China's climate policy. While China's National Climate Change Plan (2014–2020) confirmed the state's great power responsibility in climate change mitigation, it also defended the country's 'legitimate development rights and interests' (National Development and Reform Commission 2014, 4–5). Most clearly, China's Intended Nationally Determined Contribution to the United Nations Framework Convention on Climate Change published in June 2015 described China as a developing country but made no reference to great power responsibility.

Conclusion

As proponents of the English School have indicated, because climate change risks the security and functions of international society, great powers bear primary responsibility for mitigating it. While pluralists justify that responsibility by citing great powers' functional role in international society, solidarists emphasise their diplomatic responsibility to advance human values and international justice as well. Arguably, environmental stewardship is a human value, and we can therefore expect great powers to shoulder primary responsibility for mitigating climate change. Great powers should fulfil such responsibility by assuming leadership roles in international climate politics and by pursuing domestic measures to halt climate change. Due to its emerging great power status, China has played an increasingly important role in the social processes in which great power responsibilities are formulated. Because it opposes Western views on the humanitarian attributes of great power responsibility, China has been keen to position climate responsibility as an important attribute of great power responsibility.

Notes

1 By analogy, Bull (2002 [1977], 53) observes the needlessness of formulating rules requiring people to sleep or eat, 'which they may be relied upon to do', but that most societies do formulate rules prohibiting killing and stealing, 'which some of them [citizens] are likely to do, whether there are rules prohibiting this kind of behaviour or not'.

2 For examples, see Ikenberry, Mastanduno and Wohlforth (2009) and Mowle (2007). Although such a characterisation is mostly advocated by the United States,

it has also been popular in China, where politicians, academics and the general public have called for multipolarity in world affairs, even if some now increasingly advocate the *democratisation* of international relations.

3 According to Dunne (1998, 106), 'balance of power is likened to the first article of the "constitution" of international society' in the papers of the British Committee.

4 For a detailed study on the League of Nations, see Zimmern (1945).

5 For a detailed overview of those developments, see Trombetta (2008).

6 By *learning*, I do not mean that new participants simply internalise existing rules of practices but that they learn to use or seek to alter practices in ways that best serve their interests and values.

7 In using that analogy, I do not literally mean that states should be treated as thinking, feeling persons. We need not study psychology in order to understand their behaviour.

8 However, Zoellick discussed energy security, which relates closely to climate change.

9 For a detailed analysis of China's partnership diplomacy, see Su (2009, 35–41).

10 For a review of the debate, see Shambaugh (2013).

11 According to Xia (2001), those conditions are fourfold. First, if China is confident about the international security environment and international mechanisms, then it will integrate itself into international society and international governmental institutions. Second, other countries have to help China to participate in international institutions because they will benefit and doing so will generate mutual trust and facilitate cooperation. Third, the strategic balance of US–Chinese–Japanese relations should be established and maintained so that no party moves to control another. Fourth, the dispute over Taiwan should be resolved peacefully so that 'China will be more willing to play as a responsible great power in the international community' (ibid., 24–25).

12 For an in-depth study of the evolution of the concept of peaceful development, see Glaser and Medeiros (2007).

Bibliography

Aslam, Wali. 2013. *United States and Great Power Responsibility in International Society: Drones, Rendition and Invasion*. London andNew York: Routledge.

Barry, John. 1999. 'Green Politics and Intergenerational Justice: Posterity, Progress and the Environment'. In N. Ben Fairweather, Sue Elworthy, Matt Stroh & Piers H. G. Stephens (eds), *Environmental Futures.* Basingstoke: Macmillan Press, 57–72.

Buzan, Barry. 2010. 'China in international society: Is "peaceful rise" possible?'. *Chinese Journal of International Politics* 3, 5–36.

Buzan, Barry. 2004. *The United States and the Great Powers: World Politics in the Twenty-First Century.* Cambridge: Polity Press.

Brown, Chris. 2004. 'Do great powers have great responsibilities? Great powers and moral agency'. *Global Society* 18:1, 5–19.

Bukovansky, Mlada, Ian Clark, Robyn Eckersley, Richard Price, Christian Reus-Smit & Nicholas J. Wheeler. 2012. *Special Responsibilities: Global Problems and American Power.* New York: Cambridge University Press.

Bull, Hedley. 2002 [1977]. *The Anarchical Society: A Study of Order in World Politics*, 3rd edition. Basingstoke: MacMillan Press.

Charter of the United Nations. 1945. www.un.org/en/charter-united-nations/index.html.

Chinaiss. 2014. '汪洋 引领世界的是美国 中国愿加入这个体系' [Wang Yang: The world's leader is the US, China is willing to join the system]. Accessed 2 October 2016. http://observe.chinaiiss.com/html/201412/29/a753fe.html.

China Daily. 2016. 'Overseas views on NPC & CPPCC: China's great power diplomacy in 2015'. 8 March. Accessed 29 May 2017. www.chinadaily.com.cn/china/2016twosession/2016-03/08/content_23788379.htm.

Clark, Ian. 2011. *Hegemony in International Society*. New York: Oxford University Press.

Clark, Ian. 2009a. 'Towards an English School theory of hegemony'. *European Journal of International Relations* 15:2, 203–228.

Clark, Ian. 2009b. 'Bringing hegemony back in'. *International Affairs* 85:1, 23–36.

Clark, Ian. 2005. *Legitimacy in International Society*. Oxford: Oxford University Press.

Donilon, Thomas. 2013. 'Remarks by Tom Donilon, National Security Advisory to the President: "The United States and the Asia-Pacific in 2013"'. 11 March. Accessed 29 May 2017. www.whitehouse.gov/the-press-office/2013/03/11/remarks-tom-donilon-national-security-advisory-president-united-states-a

Donilon, Thomas. 2012. 'President Obama's Asia policy and upcoming trip to the region'. 15 November. Accessed 29 May 2017. http://csis.org/files/attachments/121511_Donilon_Statesmens_Forum_TS.pdf.

Donnelly, Jack. 1998. 'Human Rights: A new standard of civilization?'. *International Affairs* 74:1, 1–11.

Dunne, Tim. 1998. *Inventing international Society: A History of the English School*. London: MacMillan.

Federation of American Scientists. 1999. 'Governor Bush discusses foreign policy in speech at Ronald Reagan Library'. Accessed 2 October 2016. http://fas.org/news/usa/1999/11/pr111999_nn.htm

Fidler, David P. (2001): 'The Rreturn of the standard of civilization'. Faculty Publications, Paper 432. Accessed 14 February 2018. www.repository.law.indiana.edu/facpub/432.

Foot, Rosemary. 2001. 'Chinese power and the idea of a responsible state'. *China Journal* 45, 1–9.

Frost, Mervyn. 2003. 'Constitutive theory and moral accountability: Individuals, institutions and dispersed practices'. In Toni Erskine (ed.), *Can Institutions Have Responsibilities?*Basingstoke: Palgrave, 84–99.

Gill, Bates. 2007. 'China becoming a responsible stakeholder'. Carnegie Endowment for International Peace, 11 June. Accessed 20 February 2017. http://carnegieendowment.org/files/Bates_paper.pdf.

Glaser, Bonnie S. & Evan S. Medeiros. 2007. 'The changing ecology of foreign policy-making in China: The ascension and demise of the theory of "peaceful rise"'. *China Quarterly* 190, 291–310.

Harding, Harry. 1999. 'The uncertain future of US–China relations'. *Asia-Pacific Review* 6:1, 7–24.

Harrington, Jonathan. 2005. '"Panda diplomacy": State environmentalism, international relations and Chinese foreign policy'. In Paul G. Harris (ed.), *Confronting Environmental Change in East and Southeast Asia: Eco-Politics, Foreign Policy, and Sustainable Development*. London: United Nations University Press and Earthscan, 102–118.

Holsti, K. J. 2009. 'Theorising the causes of order: Hedley Bull's The Anarchical Society'. In Cornelia Navari (ed.), *Theorising International Society: English School Methods*. New York: Palgrave MacMillan, 125–147.

Hu, Jintao. 2012a. 'Promote win-win cooperation and build a new type of relations between major countries'. Accessed 4 March 2017. www.fmprc.gov.cn/mfa_eng/wjdt_665385/zyjh_665391/t931392.shtml.

Hu, Jintao. 2012b. 'Firmly march on the path of socialism with Chinese characteristics and strive to complete the building of a moderately prosperous society in all respects'. Accessed 4 March 2017. www.china.org.cn/china/18th_cpc_congress/2012-11/16/content_27137540.htm.

Hurrell, Andrew. 2016. 'Towards the global study of international relations'. Revista Brasileira de Política Internacional, http://dx.doi.org/10.1590/0034-7329201600208.

Hobson, John M. & Leonard Seabrooke. 2001. 'Reimagining Weber: Constructing international society and the social balance of power'. *European Journal of International Relations* 7:2, 239–274.

Ikenberry, G.John. 2009. 'Liberal internationalism 3.0: America and the dilemmas of liberal world order'. *Perspectives on Politics* 7:1, 71–87.

Ikenberry, G.John, MichaelMastanduno and William C. Wohlforth. 2009. 'Unipolarity, state behavior, and systemic consequences'. *World Politics* 61, 1–27.

Information Office of the State Council of the People's Republic of China. 2011. 'China's peaceful development'. Accessed 22 February 2018. www.gov.cn/english/official/2011-09/06/content_1941354.htm

Information Office of the State Council of the People's Republic of China. 2005. 'China's peaceful development road'. Accessed 22 February 2018. www.china.org.cn/english/2005/Dec/152669.htm

Jackson, Robert H. 2000. *The Global Covenant*. Oxford: Oxford University Press.

Jin, Canrong. 2011. *Big Power's Responsibility: China's Perspective*. Translated by Tu Xiliang. Beijing: China Renmin University Press.

Jin, Canrong. 2006. 'A new framework'. *Beijing Review*, January 20. Accessed 29 May 2017. www.bjreview.cn/EN/06-03-e/w-3.htm.

Jones, Terril Yue & Benjamin Kang Lim. 2013. 'China's new premier seeks "new type" of ties with U.S.'. Reuters, 17 March. Accessed 29 May 2017. www.reuters.com/article/2013/03/17/us-china-parliament-hacking-idUSBRE92G02320130317.

Knudsen, Tonny Brems. 2016. 'Solidarism, pluralism and fundamental institutional change'. *Cooperation and Conflict* 51:1, 102–109.

Kopra, Sanna. 2018. 'China, Great Power Management, and Climate Change: Negotiating Great Power Climate Responsibility in the UN'. In Tonny Brems Knudsen & Cornelia Navari (eds), *International Organization in the Anarchical Society: The Institutional Structure of World Order*. New York: Palgrave Macmillan.

Kopra, Sanna. 2016. 'Great power management and China's responsibility in international climate politics'. *Journal of China and International Relations* 4:1, 20–44.

Krause-Jackson, Flavia. 2013. 'Climate change's links to conflict draws UN attention'. *Bloomberg*, 15 February. Accessed 29 May 2017. www.bloomberg.com/news/articles/2013-02-15/climate-change-s-links-to-conflict-draws-un-attention

Lanteigne, Marc. 2013. *Chinese Foreign Policy*, 2nd edition. New York: Routledge.

Lee, James R. 2009. *Climate Change and Armed Conflict: Hot and Cold Wars*. London: Routledge.

National Development and Reform Commission. 2014. '国家应对气候变化规划(2014–2020年)' [National Climate Change Plan (2014–2020)]. Accessed 29 May 2017. www.ccchina.gov.cn/nDetail.aspx?newsId=49211&TId=60.

Masuda, Masayuki. 2009. 'China's search for a new foreign policy frontier: Concept and practice of "harmonious world"'. In Iida Masafumi (ed.), *China's Shift: Global Strategy of the Rising Power*. NIDS Joint Research Series no 3. Tokyo: National Institute for Defense Studies, 57–79.

Mazo, Jeffery. 2010. *Climate Conflict: How Global Warming Threatens Security and What to Do About It*. London: Routledge.

Medeiros, Evan S. 2009. *China's International Behavior*. Santa Monica, CA: RAND Corporation.

Mowle, Thomas. 2007. *Unipolar World*. Gordonsville: Palgrave Macmillan.

Neslen, Arthur. 2017. 'Bloomberg demands seat at UN climate negotiating table for cities and states'. *Climate Home News*, 11 November. Accessed 22 February 2018. www.climatechangenews.com/2017/11/11/bloomberg-demands-seat-un-climate-negotiating-table-cities-states.

Obama, Barack. 2014. 'Remarks by the president at U.N. Climate Change summit'. Accessed 22 February 2018. www.whitehouse.gov/the-press-office/2014/09/23/remarks-president-un-climate-change-summit.

Onuma, Yasuaki, 2003. 'International law in and with international politics: The functions of international law in international society'. *European Journal of International Law* 14:1, 105–139.

Pang, Zhongying. 2006. 'China, my China'. *National Interest* 83 (Spring), 9–10.

Reus-Smit, Christian. 1999. *The Moral Purpose of the State*. Princeton: Princeton University Press.

Reuters. 2017. 'China says "China responsibility theory" on North Korea has to stop'. 11 July. Accessed 22 February 2018. www.reuters.com/article/us-northkorea-missiles-china/china-says-china-responsibility-theory-on-north-korea-has-to-stop-idUSKBN19W0V6.

Roberts, J. Timmons & Bradley C. Parks. 2007. *A Climate of Injustice: Global Inequality, North-South Politics, and Climate Policy*. Cambridge, MA, and London: MIT Press.

Roosevelt, Franklin D. 1945. 'Undelivered address prepared for Jefferson Day'. Accessed 29 May 2017. http://georgiainfo.galileo.usg.edu/FDRspeeches/FDRspeech45-1.htm.

Roy, Denny. 2002. 'China and the war on terrorism'. *Orbis* 46:3, 511–521.

Rozman, Gilbert. 1999. 'China's quest for great power identity'. *Orbis* 46:3, 383–402.

Sachs, Wolfgang. 1993. 'Global ecology and the shadow of "development"'. In Wolfgang Sachs (ed.), *Global Ecology: A New Arena of Political Conflict*. London: Zed Books, 3–21.

Simpson, Gerry. 2004. *Great Powers and Outlaw States: Unequal Sovereigns in the International Legal Order*. Cambridge: Cambridge University Press.

Shambaugh, David. 2013. *China Goes Global: The Partial Power*. New York: Oxford University Press.

Shambaugh, David. 2002. 'Sino-American relations since September 11'. *Current History* 101:656, 243–249.

Steinberg, James 2009. 'Administration's vision of the U.S.-China relationship'. USC US-China Institute, 24 September. Accessed 29 May 2017. http://china.usc.edu/james-steinberg-obama-administrations-vision-us-china-relationship-september-24-2009.

Su, Hao. 2009. 'Harmonious world: The conceived international order in framework of China's foreign affair'. In Iida Masafumi (ed.), *China's Shift: Global Strategy of the Rising Power*. NIDS Joint Research Series no 3. Tokyo: National Institute for Defense Studies, 29–55.

Suzuki, Shogo. 2014. 'Journey to the West: China debates its 'great power' identity'. *Millennium*, 42:2, 632–650.

Suzuki, Shogo. 2008. 'Seeking 'legitimate' great power status in post-Cold War international society: China's and Japan's participation in UNPKO'. *International Relations* 22:1, 45–63.

Trombetta, Maria Julia. 2008. 'Environmental security and climate change: analysing the Discourse'. *Cambridge Review of International Affairs* 21: 4, 585–602.

Truman, Harry S. 1948. 'St. Patrick's Day address in New York City'. 17 March. Online by Gerhard Peters and John T. Woolley, *The American Presidency Project*. Accessed 29 May 2017. www.presidency.ucsb.edu/ws/?pid=13131.

Truman, Harry S. 1945. 'Address before a joint session of the Congress'. 16 April. Online by Gerhard Peters and John T. Woolley, *The American Presidency Project*. Accessed 29 May 2017. www.presidency.ucsb.edu/ws/index.php?pid=12282.

United Nations. 2011. 'Security Council, in statement, says "contextual information" on possible security implications of climate change important when climate impacts drive conflict'. July 20. Accessed 29 May 2017. www.un.org/press/en/2011/sc10332.doc.htm.

United Nations. 2007. 'Security Council holds first-ever debate on impact of climate change on peace, surety, hearing over 50 speakers'. 17 April. Accessed 29 May 2017. www.un.org/press/en/2007/sc9000.doc.htm.

UN General Assembly. 2009a. 'Climate change and its possible security implications', A/RES/63/281, 85th plenary meeting. 3 June.

UN General Assembly. 2009b. 'Climate change and its possible security implications: Report of the Secretary-General', A/64/350. 11 September.

UN Security Council. 2011. 'Statement by the president of the Security Council', S/PRST/2011/15. 20 July.

UN Security Council. 1992. 'Note by the president of the Security Council', S/23500. 31 January.

Wang, Yi. 2013. 'Exploring the path of major-country diplomacy with Chinese characteristics'. 27 June. Accessed 29 May 2017. www.fmprc.gov.cn/mfa_eng/wjb_663304/wjbz_663308/2461_663310/t1053908.shtml.

Watson, Adam. 1982. *Diplomacy: The Dialogue Between States.* London: Eyre Methuen.

Welzer, Harald. 2012. 'Climate wars: Why people will be killed in the twenty-first century'. Translated by Patrick Camiller. Cambridge: Polity Press.

Wenger, Étienne. 1998. *Communities of Practice: Learning, Meaning and Identity.* Cambridge: Cambridge University Press

White House. 2009. 'Joint press statement by President Obama and President Hu of China'. On file with author.

White House. 2006. 'National Security Strategy'. Accessed 29 May 2017. http://georgewbush-whitehouse.archives.gov/nsc/nss/2006.

Wight, Martin. 1999 [1946]. *Power Politics.* Edited by Hedley Bull and Carsten Holbraad. London: Leicester University Press.

Wheeler, Nicholas J. 2000. *Saving Strangers: Humanitarian Intervention in International Society.* Oxford: Oxford University Press.

World Economic Forum. 2017. 'President Xi's speech to Davos in full'. 17 January. Accessed 29 May 2017. www.weforum.org/agenda/2017/01/full-text-of-xi-jinping-keynote-at-the-world-economic-forum.

Vaha, Milla. 2015. 'Drowning under: Small island states and the right to exist'. *Journal of International Political Theory* 11:2, 206–223.

Xi, Jinping. 2015. 'Full text of Chinese president's speech at Boao Forum for Asia Annual Conference 2015'. Accessed 22 October 2016. http://english.boaoforum.org/hynew/19353.jhtml

Xi, Jinping. 2012. 'Work together for a bright future of China-US cooperative partnership'. Accessed 22 October 2016. www.fmprc.gov.cn/mfa_eng/wjdt_665385/zyjh_665391/t910351.shtml.

Xia, Liping. 2001. 'China: A responsible great power'. *Journal of Contemporary China* 10:26, 17–25.

Xinhua. 2014. 'Xi eyes more enabling int'l environment for China's peaceful development'. 30 November. Accessed 22 October 2016. http://en.people.cn/n/2014/1130/c90883-8815967.html.

Yeophantong, Pichamon. 2013. 'Governing the world: China's evolving conceptions of responsibility'. *Chinese Journal of International Politics* 6:4, 329–364.

Yu, Wanli. 2009. 'Breaking the cycle? Sino-US relations under George W. Bush administration'. In Iida Masafumi (ed.), *China's Shift: Global Strategy of the Rising Power.* NIDS Joint Research Series no 3. Tokyo: National Institute for Defense Studies, 81–98.

Zalewski, Marysia & Cynthia Enloe. 1995. 'Questions about identity in international relations'. In Ken Booth & Steve Smith (eds), *International Relations Theory Today.* Cambridge: Polity Press, 279–305.

Zimmern, Alfred. 1945. *The League of Nations and the Rule of Law 1918–1935.* London: Macmillan.

ZhangJiadong & ZhengXin. 2012. 'The role of nontraditional security in China–US relations: Common ground or contradictory arena?'. *Journal of Contemporary China* 21:76, 623–636.

Zhang, Xiaoming. 2011. 'A rising China and the normative changes in international society'. *East Asia* 28:3, 235–246.

Zheng, Yongnian. 1999. *Discovering Chinese Nationalism in China: Modernization, Identity, and International Relations.* Cambridge: Cambridge University Press.

Zoellick, Robert B. 2005. 'Whither China: From membership to responsibility? Remarks to National Committee on U.S.-China Relations'. Accessed 22 October 2016. http://2001-2009.state.gov/s/d/former/zoellick/rem/53682.htm.

5 Great power management and debate over climate responsibility

In this chapter, I briefly introduce international climate negotiations from the perspective of state environmental responsibility. In describing the evolution of the international norm of climate responsibility, with particular attention to China's contribution to that evolution, I address key events and tensions that have shaped the formation of climate responsibilities globally. As with all sorts of politics, international climate politics is not only shaped by power struggles and participants' domestic as well as international agendas but also 'about the negotiation of social identities, arguments about legitimate interests and social purposes, the formulation and execution of strategic practices, and struggles over the good and the just' (Bukovansky et al. 2012, 64). Given the complexity of both the globalised world's social structures and the phenomenon of climate change, it is impossible to trace all the relationships and processes that have prompted the emergence and formulation of international climate practices. Accordingly, while paying particular attention to notions of responsibility, I do not offer a Foucauldian genealogy of climate responsibility, for a more selective story can be told: one regarding the role of participants such as individual states, especially contributions from the European Union, and non-state organisations. Last, I do not provide a detailed theoretical or empirical explanation of how the United Nations Framework Convention on Climate Change (UNFCCC) or any other secondary institution has emerged and developed, or how it functions. Instead, I focus on the evolving notions of state environmental responsibility.[1]

China and the environmental awakening of international society

When the United Nations (UN) was founded, environmental issues were not a major concern of international society. In fact, the UN Charter did not mention the environment at all. When concern for the climate did surface at the global level, it was not due to a sole factor but to multiple critical drivers working at once, without which it would be difficult to imagine the new planetary institution that emerged. For one, prior to the politicisation of climate change, a process of environmental awakening occurred around the world, especially in the United States, in which scientists played a prominent role by identifying environmental changes, framing those changes as political problems and formulating (political) agendas both locally and globally. Consequently, after the mid-nineteenth century, the scientific community began organising international conferences devoted to addressing ways to protect nature.[2] By the late 1950s and early 1960s, evidence of increased environmental degradation had accumulated, and several books and articles on pollution, wasted resources, overstressed ecosystems and misused technology were published, including Rachel Carson's *Silent Spring* (1962), Garrett Hardin's *The Tragedy of the Commons* (1968) and the Club of Rome's *The Limits to Growth* (1972). Such texts raised public awareness and concern for environmental changes, which in turn engendered the birth of new international actors – namely, environmental non-governmental organisations (NGOs) – that expanded the environmental awakening and influenced environmental agendas locally as well as globally. In that early phase of environmental politics, the United States took the role of international leader (Falkner 2005, 590).

Held in Stockholm on 5–16 June 1972, the first international conference on environmental protection, the UN Conference on the Human Environment (UNCHE), attracted delegates from 113 countries as well as representatives from numerous NGOs, intergovernmental organisations and other specialised agencies. Because the UNCHE was the first international conference that allowed non-state actors to participate in interstate negotiations on any topic, it significantly affected the workings of international society, perhaps most notably by generating political recognition of the idea of the 'collective responsibility of nations for the quality and protection of the earth' (Caldwell 1990, 55). Ultimately, the conference produced three non-binding instruments – the Declaration of the UNCHE, an action plan and an action plan for the human environment – the last of which included 109 sets of recommendations for governments, intergovernmental agencies and NGOs. In addition, the UNCHE played an important role in establishing the UN Environmental Programme (UNEP) and generated momentum for the later development of a wide range of international environmental agreements. As a participant at the UNCHE, China, which had become a member of the UN in 1971, established a leading small group for environmental protection under the State Council earlier that year. Thus, China's first environmental policy body was formed 'in direct and urgent response to an impending international conference' even before the People's Republic of China had become a member of UN (Ross 1999, 297–298).

Although its output was mostly rhetoric, the UNCHE played an important role in the emergence of climate responsibility. It promoted environmental awareness and knowledge globally and constructed the basis of the institutional framework for the further development of international environmental law. As Charlotte Epstein (2008, 111) posits, the conference 'marked the moment where environmental groups shifted from being social movement outsiders to legitimate policy advisors' and bona fide participants of international practices. According to Melinda L. Cain, it also marked a 'major shift in the priority given to climatic issues by international organisations' (quoted in Paterson 1996, 25), consequent to which the UN organised a series of climate-related conferences.[3] In contrast to traditional international law, the Stockholm resolutions that resulted from the UNCHE articulated several recommendations regarding states' human-centric responsibilities that guided 'what governments should do in relation to their own people rather than, as in traditional international law, what a nation-state should or should not do in relation to other nation–states' (Caldwell 1990, 65). Despite its anthropocentric focus on economic and social concerns caused by environmental change, a legacy of the UNCHE was, in Lynton Keith Caldwell's (1990, 67) words, 'an enlarged and reinforced concept of environmental responsibility that had prospective bearing upon the future of international political, legal, and organizational relationships'.

Sovereignty was a major issue at the UNCHE. In the early 1970s, environmentalists began to express concerns over the clash between practices of sovereignty and global environmental problems. Sceptical of the capacity and willingness of nation-states to solve the environmental crisis, many scientists and NGO representatives in the Stockholm Environment Forum, a parallel meeting of the UNCHE, urged the establishment of supranational global governance bodies that would be loyal to the 'planet and to humanity as species' (Caldwell 1990, 62). By contrast, governments were unwilling to compromise whatsoever on national interests for the sake of environmental protection; sovereignty was especially non-negotiable for many developing countries that had achieved independence not long before the UNCHE was held. Consequently, sovereignty emerged as the foundation for the definition of *state environmental responsibility*. According to Principle 21 of the Stockholm Declaration, states 'have, in accordance with the Charter of the United Nations and the principles of international law, the sovereign right to exploit their own resources pursuant to their own environmental policies'. That right, however, was constrained by a state-centric principle of no harm; Principle 21 also declared that states have 'the responsibility to ensure that activities within their jurisdiction or control do not cause damage to the environment of other States or of areas beyond the limits of national jurisdiction' (United Nations 1972).

The Stockholm Declaration was the first piece of international law to articulate that both humans and states have environmental responsibilities. It characterised a clean environment as a fundamental right of humans, which thus bear a correlative, 'solemn responsibility to protect and improve the environment for present and future generations'. The Stockholm Declaration thus directed the greatest environmental responsibility to states and expressed that each national government is tasked with defining what it means to be *environmentally responsible* under its own legislation: 'Local and national governments will bear the greatest burden for large-scale environmental policy and action within their jurisdictions'. The document also emphasised the importance of international cooperation, which it maintained 'is also needed in order to raise resources to support the developing countries in carrying out their responsibilities in this field' (United Nations 1972).

At the UNCHE, developing countries insisted that the greatest environmental problem was the lack of development and that poverty was caused mostly by unjust practices performed by developed countries. For example, then prime minister of India, Indira Gandhi, stated:

> Many of the advanced countries of today have reached their present affluence by their domination over other races and countries, the exploitation of their own masses and own natural resources. They got a head start through sheer ruthlessness, undisturbed by feelings of compassion or by abstract theories of freedom, equality, or justice.
>
> (Quoted in Caldwell 1990, 57)

China agreed with Gandhi's assessment but went even further. From the Chinese representative, the primary reason for environmental pollution was 'the policy of plunder, aggression and war carried out by imperialist, colonialist and neo-colonialist countries, especially by the super-Powers' (UNEP 2015). Consequently, convinced that environmental degradation was a problem caused by capitalism, China suggested that communist countries did not suffer from environmental problems. Despite the isolationism of Maoist China, the nation played quite an important role at the UNCHE, which was the first UN conference it attended. Participation in the conference benefitted China by offering an opportunity for the country to re-establish ties with other states, largely because environmental issues were not viewed to be prohibitively controversial compared to, for example, nuclear testing and arms control. Since the UNCHE was the first major international conference to address environmental protection and since the field of environmental protection had no established norms, China was able to contribute to the evolution of international environmental practices from their inception. China also became a willing representative of developing countries by advocating its 10-point statement, which encapsulated the interests of all developing countries in attendance (ECO 1972). To date, despite its rapidly growing economy, China has persisted to represent itself as the leader of developing countries in international climate negotiations.

Most notably, China significantly contributed to paragraphs 2, 4 and 5 of the Stockholm Declaration (Greenfield 1979; Sohn 1973). In particular, it advanced the establishment of the relationship between the environment and economic development, albeit with emphasis on the latter: 'Economic development and social progress are necessary for the welfare of mankind and the further improvement of the environment' (ECO 1972). Among other contributions, China also highlighted the development-related needs of developing countries. With only minor changes made, Paragraph 2 of the Stockholm Declaration adopted China's proposal, which suggested that

> The conservation and improvement of the human environment is a major issue which affects the livelihood and economic development of the people throughout the world, as well as an urgent wish of the peoples of the whole world and the bounden duty of all governments.
>
> (Greenfield 1979, 219)

In associating the quality of the environment with both human wellbeing and economic development, Paragraph 2 laid the foundations for the construction of the concept of sustainable development. Perhaps even more interestingly, the paragraph indicated that each government had a general legal obligation to protect the environment. As Louis B. Sohn (1973, 440) suggests, the essence the paragraph could be rephrased to state, 'The protection and improvement of human environment is the duty of all governments'. In early drafts of the Stockholm Declaration, similar solidarist suggestions for the general responsibilities of governments were common; although states were 'rather reluctant to accept such a broad obligation of an indeterminate scope', the Chinese delegation was 'somehow able to persuade other members of the Working Group not only to accept this duty but also to put it most appropriately in the forefront of the Declaration'. As Sohn highlights, such persuasion was a 'striking accomplishment' (ibid.). Despite its active role at the UNCHE, the Chinese delegation nevertheless refused to sign the final documents because they excluded sufficiently strong socialist elements. Although it recognised the importance of environmental protection, promoting socialism was the overriding goal for Chinese diplomacy in the Maoist era (Shouqiut & Voigtsi 1993, 22).

Regarding international justice, the Stockholm Declaration sketched new ideas concerning the rights and responsibilities of developed compared to developing countries. First, the Stockholm recommendations addressed the need for additional financial resources and technology transfer from developed countries to developing countries in order to meet environmental challenges. Second, the issue of compensation was advocated as long as the developed states, especially the United States, opposed the idea (Caldwell 1990, 66). For instance, China, which supported compensation, declared, 'Each country has a right to safeguard its environment. The corporate states are discharging pollutants and the victim states have a right for compensation' (ECO 1972). Ultimately, the Stockholm Declaration introduced the idea of compensation and encouraged states to develop rules addressing the liability of states in causing environmental damage:

> States shall cooperate to develop further the international law regarding liability and compensation for the victims of pollution and other environmental damage caused by activities within the jurisdiction or control of such States to areas beyond their jurisdiction.
>
> (United Nations 1972)

Consequently, in 1978, the International Law Commission began to formulate rules regarding liability for environmental damage (Koivurova 2014, 175).

China and the UNFCCC climate negotiations

The UN Conference on Environment and Development (UNCED), held in Rio de Janeiro on 3–14 June 1992, was an enormous, unparalleled event. In attendance were the representatives of 172 states, of whom 108 were state leaders; approximately 2,400 representatives of NGOs and 17,000 participants in the parallel NGO forum; and about 10,000 on-site journalists. In particular, the extraordinary and extensive participation of NGOs at UNCED enhanced their role in later international forums as well (Porter, Welsh Brown & Chasek 2000, 69). Given the massive number of participants, the outcomes of the conference – Agenda 21, the Rio Declaration on Environment and Development, the Statement of Forest Principles, the United Nations Framework Convention on Climate Change and the United Nations Convention on Biological Diversity – can arguably be characterised as universal agreements. All the documents were guided by the concept of sustainable development, while the Rio Declaration in particular confirmed several emerging environmental norms of international society, or what lawyers might call 'principles of customary environmental law': the principle of no harm (Principle 2), the precautionary principle (Principle 15), the polluter pays principle (Principle 16), the principle of common but differentiated responsibilities (CBDR; Principle 7) and the principle of sustainable development (principles 1, 4–6 and 8). In addition, states agreed to formulate national climate programmes, establish national greenhouse gas (GHG) inventories and cooperate in the fields of adaptation, technology, science and education in response to climate change.

At UNCED, China portrayed itself as the unofficial leader of developing countries, a position for which it had begun to campaign during the previous year (Johnston 1998, 574). In June 1991, China held the Beijing Ministerial Conference on Environment and Development that, as a result, issued the Beijing Declaration, which called for international cooperation on environmental protection and sustainable development, demanded financial assistance for developing countries, asserted the right of developing countries to economic development and opposed interference in the internal affairs of developing countries (Ross 1999, 299). In essence, the Beijing Declaration included all of the principles of China's environmental diplomacy (Johnston 1998, 574). In September 1991, China and UNEP organised a Symposium on Developing Countries and International Law, which addressed a wide range of issues of interest to developing countries, including finance and technology transfer, environmental protection and human rights (Shouqiu & Voigts 1993, 26). Presumably, the chief purpose of both meetings was to formalise the collective bargaining position of developing countries in preparation for UNCED, and, as host, China emerged as the leader of the group (Economy

1998, 272). For China, the UNFCCC was a diplomatic success, for the resulting agreement affirmed all of its demands, including the basic elements of the Beijing Declaration, a strong doctrine of sovereignty, opposition to interference in internal affairs, the historic responsibility of developed countries and the development-related rights of developing countries to receive financial assistance and technology transfer. As a country excluded from Annex I of the convention (i.e. non-Annex country), China was not subjected to any emissions reduction targets and managed to fulfil its international responsibility to cooperate simply by participating in the conference. Thus, mere participation in the UNFCCC was China's contribution, and it refused to commit to making any sort of emissions cuts. Instead, it demanded that developed countries, for historical reasons, should bear all responsibility for mitigating climate change. However, non-Annex countries were nevertheless urged to publish national GHG inventories, prepare national climate programmes and contribute to research on climate change. Ultimately, then Chinese premier Li Peng ratified the UNFCCC in 1992.

The UNFCCC entered into force in 1994. At an international meeting on climate change in 1995, however, the Chinese delegation interrogated scientific findings on climate change and shared a position with the Global Climate Coalition, an US oil and coal industry lobbying organisation that presumably met with the Chinese during the meeting (Johnston 1998, 572). China continued to question the scientific findings of the Intergovernmental Panel on Climate Change at the first Conference of Parties (COP1) held in Berlin in 1995. The Chinese delegation stated that:

> Scientifically, as Article 4.1 (g) has stated, there still exist some 'uncertainties regarding the causes, effects, magnitude, and timing of climate change'. This is common knowledge. Based on this knowledge, we should be very prudent in future action.
>
> (Chinese Delegation 1995, quoted in Johnston 1998, 572)

The United States, India and Brazil shared China's stance on such rigid criticism of climate science, whereas the EU and small island states expressed their deep concern for the consequences of climate change. Nevertheless, parties managed to agree that developed countries should set quantified emissions reduction requirements within specified timeframes (e.g. 2005, 2010 and 2020) and that the targets should be written into an international protocol. Known as the Berlin Mandate, the agreement facilitated a negotiation process that resulted in the adoption of the Kyoto Protocol in 1998. At Kyoto, climate negotiations focused on two issues: how much developed countries should cut emissions and whether some sort of flexible mechanisms should be established to support the implementation of emissions reduction targets (Bodansky 2001, 36).

At Australia's insistence, the Kyoto Protocol stipulates that each country in Annex I of the protocol (i.e. developed countries) should agree on a legally binding, specific and differentiated emissions reduction target (Triggs 2001, 306). China naturally supported the differentiation between developed and developing countries. Apart from Australia, Norway and Iceland, which were allowed to increase their emissions above 1990 levels, all developed countries were urged to decrease their GHG emissions by up to 8 per cent. In line with CBDR, no emissions reduction targets were set for developing countries. The Kyoto Protocol also established reporting and verification procedures, as well as three market-based mechanisms – Clean Development Management (CDM), emissions trade and joint implementation (i.e. the Kyoto mechanisms) – in order to facilitate and monitor emissions reduction. At first, China opposed flexible mechanisms, which would have permitted developed countries to shirk their responsibility to cut GHG emissions at home while 'disregarding the living environment of people in other countries' (Harris & Yu 2009, 59). In the early 2000s, however, China gave its 'gradual if muted acceptance' of the mechanisms and began to show interest in small-scale CDM projects (ibid.). In time, it established an institutional basis for CDM projects and even became the largest beneficiary of CDM credits worldwide. Indeed, the CDM seems like an auspicious way to reduce GHG emissions in China (Heggelund, Andresen & Fritzen Buan 2010, 246–247), despite wide criticism for the measure's inability to boost emissions control and propensity for creating incentives for reluctant countries to avoid effective emissions reductions (e.g. Wara 2007; Streck & Lin 2008).

The UNFCCC defined the climate responsibility of states according to Rio Principles 2 and 7 – sovereignty and CBDR – both of which were crucial for reaching an international agreement with China and other developing countries. CBDR acknowledges that developed (i.e. Annex I countries) and developing countries (i.e. non-Annex I countries) cannot be subjected to the same standards but that state responsibility for climate change must be allocated according to national circumstances and capacities. To those ends, the UNFCCC (1992) urged developed countries to implement national climate policies 'with the aim of returning individually or jointly to their 1990 [anthropogenic emissions] levels'. However, the UNFCCC did not ask developed countries to specify such policies and thereby failed to set any legal objective or specific schedule for stabilisation. Instead, it stated that 'such a level should be achieved within a timeframe sufficient to allow ecosystems to adapt naturally to climate change, to ensure that food production is not threatened and to enable economic development to proceed in a sustainable manner'. Despite its relevance to climate change, the polluter pays principle was excluded from the UNFCCC; whereas developed countries worried about the costs of the principle, developing countries wanted to highlight the historically based responsibility of developed countries. Accordingly, CBDR was viewed to be more attractive than the polluter pays principle because the latter would have required poor polluters to pay as well (Bukovansky et al. 2012, 128).

When the United States ratified the UNFCCC in 1992, it accepted CBDR, at least in principle. Later, in 1998, US President Bill Clinton also signed the Kyoto Protocol. However, President George W. Bush refused to ratify the protocol, which he believed would hamper economic growth. In his opinion, the Kyoto Protocol was unjustified because it did impose emissions reduction obligations upon major developing emitters, including China and India (Bush 2001; Bush 2002). With its withdrawal from the Kyoto process, the United States ostensibly concluded its role as a leader of international climate politics, as well as diluted the scope of climate responsibility internationally. Nevertheless, following Russia's ratification in 2004, the Kyoto Protocol entered into force in 2005.[4] At the first Meeting of the Parties to the Kyoto Protocol in Montreal in 2005, attending states decided to establish an ad hoc working group to facilitate the negotiations of the second phase of the Kyoto Protocol (2012–2020).

China and post-Kyoto climate negotiations

The UN Conference on Climate Change in Bali (COP13) in 2007 raised great – perhaps overly great – expectations for the results of the post-Kyoto climate negotiations. States developed the Bali Action Plan, which proposed a shared vision for long-term cooperative climate action, including a long-term global goal for emissions reduction and action on mitigation, adaptation, technology and financing for after 2012, to be adopted at COP15 in Copenhagen. Because some parties central to the conference, including the United States, were not parties to the Kyoto Protocol, post-Kyoto negotiations were organised with two tracks. Notably, at COP13, China and other developing countries committed to implement nationally appropriate mitigation actions (NAMAs) in the context of sustainable development supported and enabled by measurable, reportable and verifiable (MRV) technology, financing and capacity building. Although NAMAs were optional national climate policies, not legally binding emissions reduction targets, their development marked a significant step in the negotiation process, for it had become clear that climate change mitigation would be difficult without the participation of major developing countries whose emissions had grown rapidly.[5]

Although China was no longer typical of developing countries, it continued to represent itself as one – a poor one – during post-Kyoto negotiations. It aligned its climate politics with those of all developing countries (i.e. the G77) and used rhetoric intended to entwine their interests with its own. China's government often stressed its friendship with developing countries and argued that 'China has never separated itself from other developing countries and will never do so' (Wang 2013). However, China was no longer a characteristic poor country, and its enormous GHG emissions were also known to cause significant losses and damage in poor developing countries. Its participation in the G77 became increasingly questionable, and the least-developed countries began to criticise its unwillingness to shoulder its climate responsibility.

Naturally, China's government wanted to ensure that it would not be alienated in international negotiations and thus began to expand its cooperation with other emerging powers and major emitters (e.g. Hochstetler & Milkoreit 2014; Hallding et al. 2011). Prior to COP15 in Copenhagen, China formed BASIC with Brazil, South Africa and India, which provided both a 'platform for both pushing a hard line as a collective rather than an individual' and 'cover for China by preventing it from being seen as the only recalcitrant state' (Hallding et al. 2011, 76).

China and other developing countries stated at the Copenhagen Conference in 2009 that principle of MRV should apply to internationally supported climate actions only and not voluntary, independently financed national actions. Unsurprisingly, China prioritised sovereignty by insisting that its NAMAs should not be subject to external review because they were not internationally supported (Bukovansky et al. 2012, 149). As a compromise, the Copenhagen Accord stipulated that only internationally supported actions would be subject to MRV, whereas other mitigation actions would be communicated nationally to the UNFCCC 'with provisions for international consultations and analysis under clearly defined guidelines that will ensure that national sovereignty is respected' (UNFCCC 2009). Moreover, developed countries pledged to provide developing ones with new and additional resources worth roughly USD 30 billion during 2010–2012 (i.e. fast-start finance) and to mobilise USD 100 billion annually by 2020. However, no agreement on how the funds should be mobilised in a public-to-private ratio was ever reached.

China was content with the Copenhagen Accord because it would not infringe upon its sovereignty or compromise its short-term national interests (Christoff 2010). On the international stage, however, China became a target of harsh criticism from other states, who blamed it for being 'irresponsible' and 'blocking progress' by delaying negotiations and opposing to halve GHG emissions globally by the mid-twenty-first century (Christoff 2010, 647; cf. Lynas, 2009; Porter 2009; Vidal 2009). In diplomatic terms, 'Premier Wen's absence from the larger final high-level negotiating sessions and the presence of relatively junior officials in meetings with Obama and other heads of state were read as insults' (Christoff 2010, 647).

Before the Durban Conference in 2011, China's government had opposed binding climate obligations for developing countries at every turn, and its role in efforts to mitigate climate change was largely perceived to be negative. At the Durban Conference, however, China played a more constructive part, and participating parties successfully agreed to enter another round of negotiations to formulate a climate agreement that would oblige all major emitters by 2015 and come into force by 2020. Moreover, the distinction between Annex I and non-Annex I countries was dismantled, and the proactive climate policies of developing countries were viewed to be increasingly vital to achieving the long-term goal of not raising global temperatures by more than 2°C. For its part, the European Union committed to the second commitment period under the Kyoto Protocol. Later, the 2014 Lima Accord (COP20) obliged all parties to formulate intended nationally determined contributions (INDCs) well in advance of COP21. As a result, 187 states – even exceptionally poor ones in regions of conflict (e.g. Afghanistan) – submitted INDCs to the UNFCCC that together represented roughly 95 per cent of all GHG emissions worldwide.[6] That unusually inclusive – in fact, nearly universal – participation of states demonstrated a fundamental paradigm shift in international norms of climate responsibility: that developing countries were now urged and willing to contribute to global efforts against climate change, even if CBDR was not rescinded. In other words, in response to climate change, all parties became required to 'undertake and communicate ambitious efforts' that should 'represent a progression beyond the Party's then current nationally determined contribution and reflect its highest possible ambition' (UNFCCC 2015a). Notably, China, among other countries, agreed to the paradigm change (Xi 2015).

During the Obama administration, the United States resumed its role as leader in international climate politics. Because the engagement of major emitters such as China and India was deemed crucial to agreeing upon a new international climate agreement, Barack Obama's climate diplomacy focused on obtaining their buy-in before the UN Climate Change Conference in Paris in 2015. US–Chinese cooperation in efforts towards mitigating climate change indeed improved dramatically after the bilateral meeting of President Obama and President Xi Jinping in 2013, as the *U.S.–China Joint Announcement on Climate Change* (2014) and *U.S.–China Joint Presidential Statement on Climate Change* (2015, 2016) demonstrate. A shared understanding of the climate responsibility of great powers began to evolve, as Obama (2014) explicitly acknowledged. Moreover, China's special envoy Zhang Gaoli (2014) declared at the UN Climate Summit that 'responding to climate change is what China needs to do to achieve sustainable development at home as well as to fulfil its due international obligation as a responsible major country'. At APEC in 2014, Xi (2014) also announced that as 'its overall national strength grows, China will be both capable and willing to provide more public goods for the Asia-Pacific and the world'. Although Xi did not specify what he meant by 'public goods', clean air is a typical example of goods that everyone can consume without decreasing its availability to others. Indeed, a few days later, Xi and Obama announced a historic joint climate statement in which China pledged to stem the rise of its CO_2 emissions by 2030 (White House 2014), meaning that China would no longer focus on reducing relative carbon intensity but instead lower its absolute emissions. The joint statement sent a strong signal to international society that both the United States and China acknowledged their responsibility to lead international efforts against climate change and that an international climate treaty was possible in Paris the following year.

In preparation for the Paris conference, the UNFCCC Secretariat and France worked tirelessly on the diplomatic front to ensure its success. To manage the expectations of governments around the world, for instance,

France stationed a climate change representative in all its embassies in major countries worldwide. The role of non-state actors was also crucial to paving the way for global agreement (Bailey & Tomlinson 2016, 2). The result of years of diplomacy, the Paris Agreement, adopted at COP21 in Paris in 2015 and now in effect, declares that states 'aim to reach global peaking of greenhouse gas emissions as soon as possible … and to undertake rapid reductions thereafter in accordance with best available science, so as to achieve a balance between anthropogenic emissions by sources and removals by sinks of greenhouse gases in the second half of this century' (UNFCCC 2015a). Although ultimately agreeing to strive for a carbon-neutral world, states debated that long-term goal of climate responsibility at considerable length. Eventually, the Paris Agreement came to limit 'the increase in the global average temperature to well below 2°C above pre-industrial levels and to pursue efforts to limit the temperature increase to 1.5 °C above pre-industrial levels' (ibid.), which would significantly reduce the risks and impacts of climate change. Although the goal of 1.5 °C was exalted, many analysts and NGOs did not believe that it was realistic because the Paris Agreement would lack measures effective enough to achieve it or even the 2°C goal. Since COP21 acknowledged that INDCs were insufficiently ambitious to limit the rise of global temperatures to 2°C, the Paris Agreement requires states to submit revised, increasingly ambitious INDCs by 2020 and every five years afterwards. COP21 also requested the Intergovernmental Panel on Climate Change to produce a report in 2018 to map how a global temperature increase could be limited to 1.5°C above pre-industrial levels (ibid.).

Although the Paris Agreement no longer classifies states into Annex I and non-Annex I countries, it is guided by CBDR and articulates that developed countries '*should* continue taking the lead by undertaking economy-wide absolute emission reduction targets'. Nevertheless, it also establishes a common framework for universal climate responsibility, in which it maintains that developing countries 'should continue enhancing their mitigation efforts, and *are encouraged* to move over time towards economy-wide emission reduction or limitation targets in the light of different national circumstances'. It adds moreover that developing countries need assistance to implement their national climate action plans and that their emissions peaks may come later than those of developed countries (UNFCCC 2015a, all italics added). China vigorously supported the last clause and with other BASIC countries and the Like-Minded Developing Countries on Climate Change (e.g. Argentina, Bolivia, China, Cuba, El Salvador, Ecuador, Iran, Nicaragua, Venezuela, Malaysia, Vietnam, Saudi Arabia and India) resisted legally binding emissions reductions for developing countries. In particular, both groups conceived no sub-categories between developed and developing countries, which would weaken their position in international climate negotiations. Nevertheless, China currently no longer focuses solely on the historically informed responsibility of developed countries, as Xi (2015) demonstrated in his speech to COP21 by calling for all states to 'assume more shared responsibilities for win–win outcomes'. Overall, China undoubtedly played a highly constructive role at the Paris Conference, and for the first time, China's head of state instead of its premier not only participated in international climate negotiations but also portrayed China as a determined facilitator of international climate agreement. After the conference, China's foreign ministry spokesperson Hong Lei (2015) congratulated 'China's sense of responsibility as a major country in tackling climate change'.

Despite some shortcomings – the greatest being its lack of additional mitigation contributions that would make achieving the long-term goal of limiting the rise of global temperatures to 2°C realistic – the Paris Agreement can be praised as a historic milestone for international norm of climate responsibility. As an international treaty, it obliges the states that ratify it, and though it sets no quantitative, legally binding emissions reduction obligations for any party, nor sanctions if they fail to realise their voluntary climate strategies, optimism after the conference about states' ability to fulfil their mitigation commitments was high. Perhaps the Paris Agreement's greatest strength is that it does not assign top-down obligations; on the contrary, parties to the treaty can develop voluntary, domestically appropriated mitigation plans. Its bottom-up approach thus attracted the nearly universal participation of states, which indicates not only global concern for climate change but a strong political will to combat it as well. Without a doubt, COP21 marked a crucial turning point in climate responsibility, and it now seems that climate responsibility has become an institutionalised international norm currently approaching a stage of assimilation (cf. Holsti 2004, 144–145).

China and post-Paris climate negotiations

The Paris Agreement entered into force in on 4 November 2016. More than the required 55 countries representing 55 per cent of global GHG emissions ratified the agreement in less than a year, which made it one of the fastest international agreements to ever take effect. Notably, China was among the first countries to ratify the Paris Agreement in September 2016, a decision that it publicised in a joint press conference with the United States that clearly accelerated the willingness of other states to ratify as well. Since the Paris Agreement created a global framework only of *what* states should do to limit temperature increases but not *how* they could meet that goal, COP22 in 2016 decided to adopt the so-called Paris Rulebook by 2018. Although the task was never an easy one, on the second day of COP22 the future of negotiations suddenly became even more challenging as climate sceptic Donald J. Trump was elected US president on 8 November 2016. Trump's election raised serious concerns worldwide over the US commitment to the Paris Agreement, for he had repeatedly threatened that the United States would withdraw.

Fears that Trump would vitiate the climate policies put in place by the Obama administration elevated China to a new role of leadership in international climate politics. China quickly responded to those expectations by declaring several times that it would not dilute its climate commitments despite a potential US withdrawal from the Paris Agreement (e.g. China Daily 2016). Although China has typically avoided intervening in other states' internal issues, including their elections, China's chief negotiator at the Paris Agreement, Xie Zhenhua, issued a rare comment on presidential candidate Trump's plan to abandon the Paris Agreement: a 'wise political leader should take policy stances that conform with global trends' (Wong 2016). Quickly after Trump's election, the deputy head of the Chinese climate delegation, Gou Haibo, assured the world that China would not 'change its stance on climate change' even if the United States withdrew from the Paris Agreement (China Daily 2016). Moreover, in his famous statement at the 2017 World Economic Forum in Davos, President Xi stated that '[a]ll signatories should stick to [the Paris Agreement] instead of walking away from it as this is a responsibility we must assume for future generations' (World Economic Forum 2017). Likewise, Xi's (2017a) speech at the UN Office at Geneva clarified to the world in general and to President Trump in particular that China would not dilute its commitment to the Paris Agreement. In March 2017, Chinese foreign ministry spokesman Lu Kang reportedly vowed, 'No matter how other countries' policies on climate change change, as a responsible large developing country, China's resolve, aims and policy moves in dealing with climate change will not change' (Reuters 2017).

In June 2017, Trump indeed signed an executive order to withdraw the United States from the Paris Agreement (White House 2017), which political leaders and non-state actors worldwide harshly criticised. Although Chinese representatives may not have used exceptionally strong words, they unequivocally condemned Trump's decision. Yet, Chinese leaders did not make official comments to illustrate how China would strengthen its role in international climate politics, although various commentators around the world seemed to expect China to fulfil the vacuum of leadership created by the United States. In general, China seemed to respond positively to the new expectations and, for instance, pledged to increase cooperation with the European Union in efforts to mitigate climate change. Amid such expectations, however, the European Union and China failed to issue a formal climate statement due to disagreements in trade politics in spring 2017. Nevertheless, China, the European Union and Canada held the first Ministerial on Climate Action in Montréal, Canada, in September 2017. Representatives of 34 states pursued momentum for the full implementation of the Paris Agreement and acknowledged the importance of the pre-2020 climate commitments of developed countries, an issue which would form a pivotal part of the COP decision in Bonn later (Ministerial Meeting on Climate Action Co-Chairs Summary 2017).

President Xi's speech to the 19th Chinese Communist Party Congress in October 2017 dispelled all doubts about China's willingness to lead international efforts against climate change. In his oft-quoted statement, Xi (2017b) declared, 'Taking a driving seat in international cooperation to respond to climate change, China has become an important participant, contributor, and torchbearer in the global endeavor for ecological civilization'. Although Xi (2017b) did not mention Trump by name, he nevertheless clearly signalled to his administration that '[n]o country can address alone the many challenges facing mankind; no country can afford to retreat into self-isolation'. Despite Xi's strong statement concerning China's climate leadership – altogether rare for the president – he did not specify whether China would strengthen its climate commitments or, if so, then how. By extension, China did not represent itself as a leader in that sense at the UN Climate Change Conference in Bonn the following month. By contrast, together with the Like-Minded Countries and BASIC, China resumed its pursuit of the division of developed and developing countries that the Paris Agreement had abandoned. In response, a senior Chinese climate negotiator reportedly said, 'Although we heard some different views from the developed world that we're entering into a new world without differentiation among developing and developed countries, I think that is not the truth' (Mathiesen & Li 2017).

Prior to COP23 in Bonn, both the UNEP and the World Meteorological Organization (2017) published worrying reports on the future of climate change. Above all, the reports indicated, without rapid cuts in global GHG emissions, the goal of the Paris Agreement would be impossible to achieve and a dangerous temperature hike of at least 3°C by 2100 is likely (UNEP 2017). However, neither the reports nor the fact that negotiations were chaired for the first time by a small island state, Fiji, spurred much-needed action in international climate negotiations. Nevertheless, on the second day of the conference, Syria announced it would sign the Paris Agreement, thereby making the United States the only nation in the world not party to the agreement. Since the United States cannot formally withdraw from the pact until 2020, it sent to Bonn a delegation consisting largely of the same diplomats who had represented the United States during previous rounds of negotiation. In general, the US approach demonstrated significant contingency. As before, the US priority was to oppose the division of developed and developing states, and the European Union fully supported its opposition. In addition to the official US delegation, the shadow delegation, We're Still In, representing US coalitions of states, cities and businesses opposed to Trump's anti-climate policy, launched America's Pledge to demonstrate that their coalition represents more than half of the US economy and that they should thus receive a seat at the negotiation table (Neslen 2017). Taken together, it indeed seems that limiting global temperature rise increasingly depends on the effective mitigating actions of non-state actors.

Although the Paris Agreement applies to the post-2020 era, China and other developing countries demanded that the COP23 agenda should formally discuss developed countries' pre-2020 climate commitments that they failed to implement. In particular, developed countries have failed to provide developing countries with USD 100 billion annually by 2020 as agreed upon at

COP15; moreover, the second term of the Kyoto Protocol has not entered into force because not enough countries have ratified it. Despite the decision to exclude pre-2020 actions in the Paris Rulebook, a major part of COP23 focused on pre-2020 implementations and ambitions. Importantly, it was decided that the UNFCCC would host additional stocktaking sessions in 2018 and 2019 in order to assess progress in emissions reduction as well as climate finance flows in 2018 and 2020. Such a decision resulted largely from China's climate diplomacy and indicates that China will likely be increasingly assertive in international climate negotiations.

General climate responsibilities

Henry Shue's (1993) dichotomy between the general responsibilities of all states and the special responsibilities of states with greater capabilities plays an important part in defining and distributing climate responsibility under the UNFCCC (1992), which acknowledges that 'change in the Earth's climate and its adverse effects are a common concern of humankind'. Although developing countries have repeatedly expressed reservations about accepting 'common responsibility' (Porras 1993, 28), the UNFCCC (1992) assigns general responsibilities to all parties and lists the foremost duty of each to be fulfilling the solidarist, intergenerational responsibility to 'protect the climate system for the benefit of present and future generations of humankind'. Other notable general duties include cooperation, because 'the global nature of climate change calls for the widest possible cooperation by all countries and their participation in an effective and appropriate international response', and a commitment to 'take precautionary measures to anticipate, prevent or minimize the causes of climate change and mitigate its adverse effects' (ibid.). However, those general responsibilities are limited by the principle of CBDR. Moreover, each party needs to provide information about anthropocentric GHG emissions by source and the removal of sinks, and states have to develop national climate programmes and collaborate in science, education and training, among other fields, in order to enhance the global capacity to mitigate and adapt to climate change. Notably, the UNFCCC considers sustainable development to be both a right and a responsibility of parties and ties climate responsibility to development goals by declaring that states 'have a right to, and should, promote sustainable development' (ibid.). However, because binding emissions reduction targets would impede the development-oriented objectives of developing countries, Article 3 of the UNFCCC not only highlights the historically informed responsibility of developed countries but also hints that developing ones have a right to increase their GHG emissions if such increase stems from efforts that have improved the living standards of the poor. Last, the UNFCCC also confirms that the right to sovereignty is a significant norm in international climate politics.

Largely due to US insistence, international climate negotiations have debated all states' general climate responsibilities at length. At UNCED,

China strictly opposed the proposition that protecting the climate is a 'global issue of common responsibility for all states in an indiscriminate manner' (Gao 2001, 281). Together with the G77, China resisted the idea of the policy review of a state's national development strategies and policies, which it viewed as undue interference with the internal affairs of states. As discussed earlier, China compromised on its previous position at the 2007 UN Conference on Climate Change in Bali, at which it and other developing countries committed themselves to implement NAMAs. Such commitment marked an important change in the global distribution of climate responsibility, which no longer fell exclusively to developed countries but became a general responsibility of all states. Many developing countries submitted their NAMAs by 2012, and many indeed pledged to undertake action comparable to or even more ambitious than those of developed countries (e.g. Held, Roger & Nag 2013). Another sign of the erosion of the former distinction of developed and developing countries came with the Paris Agreement, which establishes a framework of transparency with a common binding commitment for all states to submit a 'national inventory report of anthropogenic emissions by sources and removals by sinks of greenhouse gases' and to provide information 'necessary to track progress made in implementing and achieving' their national nationally determined mitigation and adaptation goals (UNFCCC 2015a). Reaching such an agreement required China, which had previously regarded reporting obligations as a violation of its sovereignty, to compromise.

In addition to mitigating climate change, adapting to it plays an important role in climate politics. Apart from developed countries' special responsibility to provide developing countries with the funding, equipment and know-how necessary to adapt to climate change, adaptation has not been a highly contested issue in international climate negotiations. Previously, adaptation was not discussed at length in international climate negotiations because doing so would have been viewed as avoiding the responsibility to reduce emissions. The UNFCCC (1992) does not clearly define *adaptation* but acknowledges that all states have a general responsibility to 'take precautionary measures to anticipate, prevent or minimize the causes of climate change and mitigate its adverse effects'. However, it is becoming increasingly clear that states have failed to prevent climate change, and the climate system continues to change regardless of states' actions. Even if a highly ambitious global treaty is reached, climate change will continue to pose significant threats to human security by, for example, increasing the frequency and intensity of extreme weather events, enhancing the spread of infectious diseases and harming food production. Therefore, adaptation has now attracted greater attention from politicians and academics alike. In particular, an extensive body of literature addresses resilience, a concept which social scientists have adopted from ecology and other fields in the natural sciences. Nevertheless, there is no common understanding of what sort of concrete policies and actions adaptation should involve. Due to countries' diverse national circumstances such as their geographical, demographic and socio-economic characteristics, the risks posed by climate change vary from country to country, and the potential for and reasonable means of adaptation are therefore diverse as well. When defining *adaptation*, Barry Smit et al. (2000) suggest considering at least four dimensions: what needs to be adapted to, who or what needs to adapt, how that adaptation should transpire and, ultimately, the extent to which the adaptation is good or appropriate. In general, adaptation can be proactive or reactive, although it invariably includes, for example, the risk assessments of health, food security, agricultural, environmental, economic and disaster management as well as strategies to alleviate and respond to those risks. Ultimately, adaptation is a domestic matter, and responsibility for developing and implementing national adaptation plans falls to the governments of states. In short, adaptation can thus be viewed as the general national responsibility of all states.

Special climate responsibilities

According to the UNFCCC, CBDR stands as a cornerstone of international climate responsibility by holding that developed countries have special responsibility to respond to climate change. Although developed countries have not caused climate change intentionally, their means of industrialisation have made them affluent at the expense of a clean environment and impoverished developing countries in the process. Developing countries not only suffer from the impacts of climate change but also lack sufficient resources to contribute to mitigating climate change. Developed countries therefore have a greater responsibility to help developing countries to meet basic standards of living, as well as a positive responsibility not only to reduce emissions but also to support developing countries' efforts to cope with climate change. Although developed countries generally accept CBDR, their agreement on their special responsibilities related to climate change remains complicated. In particular, developed states have disputed two issues: how much emissions the United States and other industrialised countries should be ordered to cut and how much (financial) assistance developed countries should offer developing countries in order to increase their capabilities to respond to climate change.

Doubtlessly, ambitious emissions reductions are the most essential aspect of climate responsibility; if no effective action is taken, then the climate will change at an increasingly rapid rate. Although the UNFCCC acknowledged that dynamic, it did not set quantitative emissions reductions requirements for any party due to US refusal to accept an emissions reduction target. In accordance with CBDR, the UNFCCC (1992) stated that developed countries have to take the 'lead in combating climate change and the adverse effects thereof' and that Annex I countries in particular should 'adopt national policies and take corresponding measures on the mitigation of climate change, by limiting its anthropogenic emissions of GHGs and protecting and enhancing its greenhouse gas sinks and reservoirs'. However, it did not require any country to achieve that target. The Kyoto Protocol, however, put

CBDR into practice. In contrast to the UNFCCC, which encourages developed countries to decrease their emissions, the Kyoto Protocol set quantitative requirements for them and ordered developed countries to cut their overall GHG emissions by 5 per cent compared to 1990 levels during the first five-year period (i.e. 2008–2012).

CBDR and the unambitious notions of climate responsibility promoted by the United States dominated post-Kyoto climate negotiations as well. The George W. Bush administration pursued the replacement of CBDR with a so-called 'mutual-burden sharing' approach that denied the historically informed responsibility of developed countries and focused instead on reducing future emissions intensity via technological innovation (Bukovansky et al. 2012, 144). By contrast, China opposed any emissions reduction targets for developing countries and emphasised the differences between the luxury emissions of developed countries and the survival emissions of developing ones. Following the US presidential election of Barack Obama in 2008, new hopes for greater US climate responsibility were aroused. Unlike the Bush administration, the Obama administration acknowledged the special climate responsibilities of the United States based on historically informed responsibility and material capabilities (e.g. Obama 2014; Obama 2015). However, Obama's commitments did not translate into sufficiently ambitious domestic climate policies. In particular, due to resistance in the US Congress, the United States could not commit to a legally binding emissions reduction target at the Paris Conference in 2015. Therefore, the Paris Agreement does not specify a legally binding emissions reduction target for any state but is based on states' voluntary, nationally determined contributions instead.

Although CBDR maintains that developed countries have a special responsibility to assist developing ones to respond to climate change, implementing that responsibility has been disputed for decades. As Robert Jackson (1996, 185–186) indicates, CBDR was nevertheless a 'significant step in the development of normative international relations' because the special responsibility was not characterised as 'aid' but as a 'responsibility'. In effect, technology transfer and other assistance cannot be regarded as charity because 'everyone, including developed countries, will benefit from such transfers' (ibid). In practice, the UNFCCC (1992) urged developed countries to provide developing ones with 'new and additional financial resources' and technology transfer to respond to climate change. It acknowledged:

> The extent to which developing country Parties will effectively implement their commitments under the Convention will depend on the effective implementation by developed country Parties of their commitments under the Convention related to financial resources and transfer of technology and will take fully into account that economic and social development and poverty eradication are the first and overriding priorities of the developing country Parties.
>
> (UNFCCC 1992)

This paragraph has formed a cornerstone of China's position on international climate negotiations. Over the years, states have established various financial mechanisms to coordinate and implement the special responsibility of developed countries to assist policies for mitigating climate change and the actions of developing countries. As a case in point, the UNFCCC established a financial mechanism to offer funds to developing countries; initially, such assistance was channelled through the Global Environmental Facility, either directly or through two climate funds, the Least Developed Country Fund and the Special Climate Change Fund. Later, the Adaptation Fund was established in 2001 in order to aid developing countries of the Kyoto Protocol to finance their concrete adaptation projects. At the 2009 UN Conference of Climate Change in Copenhagen, developed countries made significant decisions on climate finance. Furthermore, held in Cancún in 2010, COP16 established the Green Climate Fund and adopted many other concrete institution-building decisions related to adaptation (i.e. the Cancun Adaptation Framework), technology (i.e. the Technology Mechanism) and forests (i.e. REDD+), for instance. On the one hand, UN climate funds have helped developing countries to begin to meet the challenges posed by climate change (Nakhooda & Norman 2014), although developed countries have not yet fulfilled their promises of additional resources. On the other, many bilateral and multilateral funds, including the Climate Investment Funds, have also been established as a result of growing disaffection with UN funds, as a list provided by the UNFCCC (2011) shows.

China has criticised international climate negotiations for overly focusing on mitigating climate change and paying too little attention to efforts to adapt to it and thus failing to 'meet the actual needs of developing countries, in particular the least developed countries and small island countries' (National Development and Reform Commission 2008). Since adaptation is 'an essential component in the framework of sustainable development to address climate change', China demands that developed countries provide developing countries with the technological and financial support to develop their adaptation capacity (ibid.).

The Paris Agreement requires developed countries to 'provide financial resources to assist developing country Parties with respect to both mitigation and adaptation' (UNFCCC 2015a). Such funds can be mobilised 'from a wide variety of sources, instruments and channels, noting the significant role of public funds' and the 'mobilization of climate finance should represent a progression beyond previous efforts' (ibid.). However, the agreement does not pinpoint how much funding developed countries should provide because such a binding commitment would require approval from the US Congress. Nevertheless, countries at COP21 decided that in addition to the current goal of an annual USD 100 billion, developed countries will extend their financial assistance with a so-called 'new collective quantified goal' for the period after 2025 (ibid.), by which funds will be allocated through the UNFCCC's Financial Mechanism.

In addition to developed countries' special responsibility to assist developing countries to mitigate GHG emissions and adapt to climate change, developing countries have increasingly urged them to also bear a third sort of special responsibility: compensation for the losses and damage that climate change causes in developing countries in general and small island states in particular. Such efforts recall the ambitious objectives of the UNCHE to develop a compensation framework for environmental harm. The Rio Conference, unfortunately, diluted the original idea; although it promoted 'expeditious' international cooperation in developing 'international law regarding liability and compensation for adverse effects of environmental damage' (Rio Declaration on Environment and Development 1992), it subordinated liability and compensation to states' national legislation. According to the Rio Declaration (1992), 'States shall develop national law regarding liability and compensation for the victims of pollution and other environmental damage.' Likewise, the International Law Commission issued preventative rules to prevent transboundary environmental damage in 1999, which differs starkly from the original idea of rules regarding liability to facilitate compensation for victims of environmental damage (Koivurova 2014, 175). Partly in response, the Warsaw International Mechanism for Loss and Damage associated with Climate Change Impacts (i.e. Loss and Damage Mechanism) was established in 2013 to enhance 'knowledge and understanding of comprehensive risk management approaches to address loss and damage associated with the adverse effects of climate change, including slow onset impacts' (UNFCCC 2014).

As a result of intensive debate, *loss and damage* won its own article in the Paris Agreement. Notably, parties at COP21 requested that the Loss and Damage Mechanism not only 'establish a clearinghouse for risk transfer that serves as a repository for information on insurance and risk transfer' but also 'develop recommendations for integrated approaches to avert, minimise and address displacement related to the adverse impacts of climate change' (UNFCCC 2015a). In so doing, the parties acknowledged a special responsibility particularly important to realising international justice: that developed countries have a moral duty to assist developing countries to manage, for example, damages caused by floods and extreme weather events. That responsibility is an altruistic one, for developed countries themselves do not benefit from such assistance. Essentially, loss and damage assistance therefore differs from mitigation and adaptation assistance, which also advances developed countries' interests in terms of global emissions reductions and the creation of business opportunities, among other ends. Nevertheless, since parties at COP21 noted in response to US demands that the ratification of the Paris Agreement would not 'involve or provide a basis for any liability or compensation' (ibid.), the principle of no harm has persisted as the critical rule for climate change damage, one that applies to all states, even though CBDR continues to underscore the enhanced capacities of developed countries to prevent environmental harm in practice (Voigt 2008, 17).

The UNFCCC has not articulated the special responsibilities for great powers but such responsibilities have been assigned to developed countries in general. However, that situation does not mean that all developed states are expected to shoulder similar responsibilities in practice. Small developed countries such as Portugal and Cyprus are not under the same pressure as the United States or the United Kingdom to take action, largely given their differentiated material capabilities and national circumstances, and that difference has indeed been a major question in the negotiations process (Bukovansky et al. 2012, 131). Another reason is that great powers are expected to play the role of leader in global governance, meaning that, whether or not climate change is governed in the UN Security Council, great powers have a responsibility to lead global efforts to combat climate change. Accordingly, the United States has the greatest responsibility to lead, although China cannot avoid its own level of global responsibility. In 2013, the two countries indeed recognised the important role of US–Chinese climate cooperation as a 'powerful example that can inspire the world' (White House 2013). In September 2014, President Obama (2014) linked climate responsibility and great power status by declaring that the United States and China 'have a special responsibility to lead' international action to mitigate climate change because that is 'what big nations have to do'. However, the recognition of great powers' special climate responsibility has not transformed into the acceptance of legally binding emissions reductions targets within the UNFCCC, largely because both China and the United States have resisted legal obligations and emphasised nationally determined climate strategies instead.

The special climate responsibilities of developed countries are not accompanied by corresponding special rights or privileges, nor do great powers have any privileges under the UNFCCC. That dynamic likely explains in part why the United States, which normally joins international treaties in case they include exemptions, failed to ratify the Kyoto Protocol (Chalecki 2009, 152). Flexible mechanisms, established three years after CBDR was accepted in Rio (Bukovansky et al. 2012, 130), undoubtedly facilitate the execution of the special climate responsibilities of developed countries; however, they can hardly be viewed as privileges.

Great climate power clubs

The UNFCCC is not the only forum for defining and institutionalising climate responsibility. Because multilateral climate negotiations have been troublesome and proceeded at a crawl, many observers have suggested that they should be replaced with so-called 'minilateralism' (cf. Falkner 2015). As Moisés Naím (2009, 135) suggests, parties 'should bring to the table the smallest possible number of countries needed to have the largest possible impact on solving a particular problem'. For climate negotiations, minilateralism would mean smaller forums of major emitters only. Naím proposes that 20 is the 'magic number' because 20 major polluter states account for 75 per cent of the world's GHG emissions (ibid.). Such intergovernmental forums representing the world's major emitters have been established over the years and include the Asia-Pacific Partnership on Clean Development and Climate (APP) and the Major Economies Forum on Energy and Climate (MEF), for instance.

Dissatisfied with the Kyoto Protocol, in 2006 President George W. Bush initiated APP, in which a pro-market coalition of Australia, Canada, China, India, Japan, South Korea and the United States worked 'to meet goals for energy security, national air pollution reduction, and climate change in ways that promote sustainable economic growth and poverty reduction' and 'focused on expanding investment and trade in cleaner energy technologies, goods and services in key market sectors' (Asia-Pacific Partnership on Clean Development and Climate 2011). Consequently, APP focused on reducing future emissions intensity by way of technological innovation and transfer and did not bother addressing 'questions of equity, historical responsibility, capabilities, or the ethical implication of cumulative per capita emissions' (Bukovansky et al. 2012, 143–144). Completed in 2011, APP's win–win approach did not require significant costs or other sacrifices from any party, and participation did not pose any risk, either.

MEF, initiated by President Obama and launched on 28 March 2009, did not attempt to replace the UNFCCC but aimed at spurring UN negotiations in Copenhagen. In particular, it intended to facilitate a candid dialogue among major developed and developing economies, help to generate the political leadership necessary to achieve a successful outcome at the annual UN climate negotiations and advance the exploration of concrete initiatives and joint ventures to increase the supply of clean energy while reducing GHG emissions. The 17 participants of MEF have included Australia, Brazil, Canada, China, the European Union, France, Germany, India, Indonesia, Italy, Japan, Korea, Mexico, Russia, South Africa, the United Kingdom and the United States (Major Economies Forum on Energy and Climate 2015). During the Trump administration, however, no new MEF meeting has been scheduled.

In line with its growing global role, China has established its own clubs in international politics. Many of those organisations and mechanisms run parallel to their US-led counterparts (e.g. BRICS vs. the G8 and the Shanghai Cooperation Organisation vs. the North Atlantic Treaty Organization). Various Chinese initiatives have addressed climate change and put forward proposals to enhance green development. At the first One Belt, One Road forum in Beijing in 2017, for instance, President Xi (2017c) pledged to establish an 'international coalition for green development on the Belt and Road' and to 'provide support to related countries in adapting to climate change'. In practice, however, the Belt and Road concentrates on infrastructure and fossil fuel energy projects, and its impacts on climate change are likely to be negative. By contrast, China's collaboration with the European Union and Canada under the Ministerial on Climate Action may pave the way for the emergence of a new type of great climate power club.

Despite their potential to increase the willingness and capacity of major emitters to reduce emissions, minilateral forums can be criticised from an ethical perspective. Above all, they do not consider the viewpoints of developing countries, which are most vulnerable to the effects of climate change. In response, Robyn Eckersley (2012) has proposed embracing inclusive minilateralism in the form of a global climate council. Procedurally, such a council would be based on common but differentiated representation by the most capable (i.e. leading industrialised countries), the most responsible (i.e. the greatest emitters of historical, aggregate and forecasted GHG emissions) and the most vulnerable (i.e. developing countries that suffer the most from climate change and have the least capacity to respond). All three groups would be included in a council composed of the United States, the European Union, China, India, Russia, Japan, Brazil, South Korea, Mexico plus representatives of the Alliance of Small Island States, the African Group and the Least Developed Countries. That group of 12 would represent approximately 70 per cent of all emissions worldwide and roughly 70 per cent of the world's population (Eckersley 2012, 35–36). Such a composition, however, would wholly exclude civil society, whose role in environmental negotiations has been essential. Environmental NGOs, for example, have played a critical role in identifying issues, setting agendas, facilitating education and communication about the environment, formatting policy, democratising environmental decision making, pursuing normative development, organising political pressure for states, international organisations and companies and monitoring the implementation of environmental standards.

From the perspective of climate responsibility, there is no univocal argument either for or against minilateralism in climate negotiations. On the one hand, minilateralism could enhance political dialogue, identify new solutions to climate governance and help to more explicitly define the special responsibilities of participants of a climate power club, even if such a club would motivate or pressure participants to take more ambitious action to reduce emissions. On the other, minilateral groups would not generate the much-needed political will to reduce emissions globally. In particular, they erode the legitimacy of the UNFCCC and would not encourage cosmopolitan climate responsibility. Therefore, minilateral clubs would work best when made complementary to multilateral climate practices.

Conclusion

Whether or not China was prepared, Donald Trump's election as US president rapidly changed China's role in international climate politics. Above all, the world began to expect China to shoulder increasingly more responsibility to mitigate climate change. However, international expectations of China's leadership could prove to be overly optimistic, and as a result, no state could assume leadership in international climate politics. Nevertheless,

China has convinced the world that it will uphold the Paris Agreement, and from a pluralist perspective, China no doubt aims to fulfil its responsibility to uphold international law. Despite the content of speeches delivered by its leaders, China has not signalled that its approach to international climate negotiations will be more solidarist than before, despite that approach being necessary to spurring more ambitious emissions reduction pledges required to meet the goals of the Paris Agreement around the world. If China uses its increasing leverage in international climate politics to restore explicit bifurcation in the Paris Rulebook, then a remarkably troublesome future for international climate negotiations during the next few years is in store. At the same time, Trump's decision to withdraw from the Paris Agreement opened up possibilities for other sorts of global leadership that could be far more solidarist than ever before: the emergence of the We're Still In coalition in 2017 indicates the growing role of non-state actors in international relations, which, in turn, raises the question of whether non-state actors could intervene and perform climate responsibility instead of states, whose negotiations have proven to be insufficiently effective. For the English School, that possibility raises other interesting questions about institutional change and the role of civil society actors in international society.

Notes

1 For comprehensive accounts of the development of climate regime, see Yamin and Depledge (2004) and Schiele (2014), and for developments of international environmental law, see Bodansky (2010) and Koivurova (2014).

2 According to Caldwell (1990, 41), the first important conferences included the International Congress for the Protection of Nature in Paris in 1909, the International Congress on the Protection of Flora, Fauna, and Natural Sites and Monuments in Paris in 1923, the International Congress for Study and Protection of Birds in Geneva in 1927 and the Second International Congress for the Protection of Nature in Paris in 1931.

3 Such conferences included the UN World Food Conference in 1974, the UN Water Conference in 1976 and the UN Desertification Conference in 1977, all of which identified climate change as a central concern.

4 Participants agreed that the Kyoto Protocol would enter into force when at least 55 countries representing at least 55% of greenhouse gas emissions produced by developed countries ratified it.

5 See, for example, Vihma, Mulugetta and Karlsson-Vinkhuyzen (2011) for developing countries' role in international climate negotiations and Hallding et al. (2011) for the cooperation of emerging powers in efforts to mitigate climate change.

6 For a summary of all INDCs, see UNFCCC (2015b) and www.carbon-pulse.com/indcs/.

Bibliography

Asia-Pacific Partnership on Clean Development and Climate. 2011. 'Welcome to the Asia-Pacific Partnership on Clean Development and Climate'. Accessed 20 October 2016. www.asiapacificpartnership.org/english/Default.aspx.

Bailey, Rob & Shane Tomlinson. 2016. 'Post-Paris: Taking forward the global climate change deal'. Energy, Environment and Resources Briefing, Chatman House.

Bodansky, Daniel. 2010. *The Art and Craft of International Environmental Law.* Cambridge, MA: Harvard University Press.

Bodansky, Daniel. 2001. 'History of the global climate change regime'. In Urs Luterbacher & Detlef F. Sprinz (eds), *International Relations and Global Climate Change.* Cambridge andLondon: MIT Press, 23–40.

Bukovansky, Mlada, Ian Clark, Robyn Eckersley, Richard Price, Christian Reus-Smit & Nicholas J. Wheeler. 2012. *Special Responsibilities. Global Problems and American Power.* New York: Cambridge University Press.

Bush, George W. 2002. 'President announces clear skies & global climate change initiatives'. Accessed 20 October 2016. http://georgewbush-whitehouse.archives.gov/news/releases/2002/02/20020214-5.html.

Bush, George W. 2001. 'Text of a letter from the president to senators Hagel, Helms, Craig, and Roberts'. Accessed 20 October 2016. http://georgewbush-whitehouse.archives.gov/news/releases/2001/03/20010314.html.

Caldwell, Lynton Keith. 1990. *International Environmental Policy: Emergence and Dimensions*, 2nd edition. Durham andLondon: Duke University Press.

Chalecki, Elizabeth L. 2009. 'Exceptionalism as foreign policy: US climate change policy and an emerging norm of compliance'. In Paul G. Harris (ed.), *Climate Change and Foreign Policy.* New York: Routledge, 148–161.

China Daily. 2016. 'Climate meet stays on track'. 11 November. Accessed 20 February 2017. www.chinadaily.com.cn/china/2016-11/11/content_27343097.htm.

Christoff, Peter. 2010. 'Cold climate in Copenhagen: China and the United States at COP15'. *Environmental Politics* 19:4, 637–656.

Eckersley, Robyn. 2012. 'Moving forward in the climate negotiations: Multilateralism or minilateralism?'. *Global Environmental Politics* 12:2, 24–42.

ECO. 1972. 'Chinese Declaration'. Accessed 20 October 2016. www.government.se/content/1/c6/18/49/17/b2032147.pdf.

Economy, Elizabeth. 1998. 'China's environmental diplomacy'. In Samuel S. Kim (ed.), *Chinese Foreign Policy Faces the New Millennium*, 4th edition. Boulder, CO: Westview Press, 264–283.

Epstein, Charlotte. 2008. *The Power of Words in International Relations.* Cambridge: MIT Press.

Falkner, Robert. 2015. 'A minilateral solution for global climate change? On bargaining efficiency, club benefits and international legitimacy'. Centre for Climate Change Economics and Policy Working Paper no 222/Grantham Research Institute on Climate Change and the Environment Working Paper no 197.

Falkner, Robert. 2005. 'American hegemony and the global environment'. *International Studies Review* 7, 585–599.

Gao, Zhiguo. 2001. 'The Kyoto Protocol and the international energy industry: Legal and economic implications of implementation. The Chinese perspective'. In Peter D. Cameron & Donald Zillman (eds), *Kyoto: Form Principles to Practice.* Hague: Kluwer Law International, 275–287.

Greenfield, Jeanette. 1979. *China and the Law of the Sea, Air, and Environment.* Alphen aan den Rijn: Sijthoff & Noordhoff.

HalldingKarl, Marie Olsson, Aaron Atteridge, Antto Vihma, Marcus Carson & Mikael Román. 2011. *Together Alone: BASIC Countries and the Climate Change Conundrum.* Copenhagen: Tema Nord.

Harris, Paul G. & Hongyuan Yu. 2009. 'Climate change in Chinese foreign policy: Internal and external responses'. In Paul G. Harris (ed.), *Climate Change and Foreign Policy.* New York: Routledge, 53–67.

Heggelund, Gørild, Steinar Andresen & Inga Fritzen Buan. 2010. 'Chinese climate policy: Domestic Priorities, Foreign Policy, and Emerging Implementation'. In Kathryn Harrison & Lisa McIntosh Sundstrom (eds), *Global Commons, Domestic Decisions.* Cambridge, MA: MIT Press, 229–259.

Held, David, Charles Roger & Eva-Maria Nag. 2013. *Climate Governance in the Developing World.* Cambridge: Polity Press.

Hochstetler, Kathryn & Manjana Milkoreit. 2014. 'Emerging powers in the climate negotiations: Shifting identity conceptions'. *Political Research Quarterly* 67:1, 224–235.

Holsti, K. J. 2004. *Taming the Sovereigns: Institutional Change in International Politics.* Cambridge: Cambridge University Press.

Hong, Lei. 2015. 'Foreign ministry spokesperson Hong Lei's remarks on the outcomes of the Paris climate conference'. Accessed 24 September 2016. www.fmprc.gov.cn/mfa_eng/xwfw_665399/s2510_665401/t1323918.shtml

Jackson, Robert H. 1996. 'Can international society be green?'. In Rick Fawn & Jeremy Larkins (eds), *International Society after the Cold War: Anarchy and Order Reconsidered.* Basingstoke: MacMillan Press.

Johnston, Alistair. I. 1998: 'China and international institutions: A decision rule analysis'. In Michael B. McElroy, Chris P. Nielson & Peter Lydon (eds), *Energizing China. Reconciling Environmental Protection and Economic Growth.* Cambridge, MA: Harvard University Press, 555–599.

Koivurova, Timo. 2014. *Introduction to International Environmental Law.* New York: Routledge.

Lynas, Mark. 2009. 'How do I know China wrecked the Copenhagen Deal? I was in the room'. *The Guardian*, 22 December. Accessed 26 February 2018. www.theguardian.com/environment/2009/dec/22/copenhagen-climate-change-mark-lynas.

Major Economies Forum on Energy and Climate. 2015. 'Major Economies Forum'. Accessed 3 June 2016. www.majoreconomiesforum.org.

Mathiesen, Karl & Li Jing. 2017. 'Trump's 'top priority' at climate talks: Protecting an Obama legacy'. *Climate Home News*, 14 November. Accessed 20 February 2018. www.climatechangenews.com/2017/11/14/trump-priority-climate-talks-no-soft-option-china.

'Ministerial meeting on climate action co-chairs summary'. 2017. Accessed 20 February 2018. www.canada.ca/en/environment-climate-change/news/2017/09/ministerial_meetingonclimateaction.html.

Naím, Moisés. 2009. 'Minilateralism'. *Foreign Policy* 173 (July/August), 135–136.

Nakhooda, Smita & Marigold Norman. 2014. 'Climate finance – is it making a difference? A review of the effectiveness of Multilateral Climate Funds'. London: Overseas Development Institute.

National Development and Reform Commission. 2008. 'Mr. Xie Zhenhua unveiled "Initiative By China on Enhancement of Developing Countries' Adaptation Capacity"'. Accessed 2 August 2014. http://en.ccchina.gov.cn/Detail.aspx?newsId=38737&TId=107

Neslen, Arthur. 2017. 'Bloomberg demands seat at UN climate negotiating table for cities and states'. *Climate Home News*, 11 November. Accessed 20 February 2018. www.climatechangenews.com/2017/11/11/bloomberg-demands-seat-un-climate-negotiating-table-cities-states/

Obama, Barack. 2015. 'Remarks by President Obama at the First Session of COP21'. Accessed 20 February 2018. www.whitehouse.gov/the-press-office/2015/11/30/remarks-president-obama-first-session-cop21.

Obama, Barack. 2014. 'Remarks by the President at U.N. Climate Change summit'. Accessed 22 February 2018. www.whitehouse.gov/the-press-office/2014/09/23/remarks-president-un-climate-change-summit.

Paterson, Matthew. 1996. *Global Warming and Global Politics.* London: Routledge.

Porras, Ileana. 1993. 'The Rio Declaration: A new basis for international cooperation'. In Philippe Sands (ed.), *Greening International Law.* London: Earthscan Publications, 20–33.

Porter, Andrew. 2009. 'China and America to blame for Copenhagen failure, says Brown'. *The Telegraph*, 21 December. Accessed 26 February 2018. www.telegraph.co.uk/news/politics/6859567/Gordon-Brown-Copenhagen-China.html.

Porter, Gareth, Janet Welsh Brown & Pamela S. Chasek. 2000. *Global Environmental Politics*, 3rd edition. Boulder: Westview Press.

Reuters. 2017. 'China says committed to Paris accord as Trump undoes U.S. climate policy', 29 March. Accessed 22 February 2018. www.reuters.com/article/us-usa-trump-energy-china/china-says-committed-to-paris-accord-as-trump-undoes-u-s-climate-policy-idUSKBN1700RU.

Rio Declaration on Environment and Development. 1992. Accessed 22 February 2018. www.un.org/documents/ga/conf151/aconf15126-1annex1.htm.

Ross, Lester. 1999. 'China and environmental protection'. In Elizabeth Economy & Michel Oksenberg (eds), *China Joins the World: Progress and Prospects.* New York: Council on Foreign Relations Press, 296–325.

Schiele, Simone. 2014. *Evolution of International Environmental Regimes: The Case of Climate Change.* Cambridge: Cambridge University Press.

Shue, Henry. 1993. 'Subsistence emissions and luxury emissions'. *Law & Policy* 15:1, 39–59.

Shouqiu, Cai & Mark Voigts. 1993. 'The development of China's environmental diplomacy'. *Pacific Rim Law & Policy Journal* 3, 17–42.

Smit, Barry, Ian Burton, Richard J. T. Klein & J. Wandel. 2000. 'An anatomy of adaptation to climate change and variability'. *Climatic Change* 45:1, 223–251.

Sohn, Louis B. 1973. 'The Stockholm Declaration on the Human Environment'. *The Harvard International Law Journal* 14:3, 422–515.

Streck, Charlotte & Jolene Lin. 2008. 'Making markets work: A review of CDM performance and the need for reform'. *European Journal of International Law* 19:2, 409–442.

Triggs, Gillian. 2001. 'The Kyoto Protocol and the energy industry: Australia and the Asian Pacific'. In Peter D. Cameron & Donald Zillman (eds), *Kyoto: Form Principles to Practice.* The Hague: Kluwer Law International, 299–324.

United Nations. 1972. 'Declaration of the United Nations Conference on the Human Environment'. Accessed 3 March 2017. www.un-documents.net/unchedec.htm.

UNEP (United Nations Environment Programme). 2017. 'Emissions gap report'. Accessed 20 February 2018. www.unenvironment.org/resources/emissions-gap-report.

UNEP (United Nations Environment Programme). 2015. 'Working group on the Declaration on the Human Environment'. Accessed 14 March 2017. www.unep.org/documents.multilingual/default.asp?DocumentID=97&ArticleID=1529&l=en.

UNFCCC (United Nations Framework Convention on Climate Change). 2015a. 'Adoption of the Paris Agreement: Proposal by the president', 12 December.

Accessed 14 March 2017. http://unfccc.int/documentation/documents/advanced_search/items/6911.php?priref=600008831.

UNFCCC (United Nations Framework Convention on Climate Change). 2015b. 'Synthesis report on the aggregate effect of the intended nationally determined contributions'. Accessed 14 March 2017. http://unfccc.int/focus/indc_portal/items/9240.php.

UNFCCC (United Nations Framework Convention on Climate Change). 2014. 'Warsaw international mechanism for loss and damage associated with climate change impacts.' Accessed 14 March 2017. http://unfccc.int/adaptation/workstreams/loss_and_damage/items/8134.php.

UNFCCC (United Nations Framework Convention on Climate Change). 2011. 'Multilateral and bilateral funding sources'. Accessed 24 March 2017. http://unfccc.int/cooperation_and_support/financial_mechanism/bilateral_and_multilateral_funding/items/2822.php.

UNFCCC (United Nations Framework Convention on Climate Change). 2009. 'Copenhagen Accord: Proposal by the president'. Accessed 24 March 2017. unfccc.int/resource/docs/2009/cop15/eng/l07.pdf.

UNFCCC (United Nations Framework Convention on Climate Change). 1992. 'United Nations Framework Convention on Climate Change'. Accessed 3 March 2017. http://unfccc.int/files/essential_background/background_publications_htmlpdf/application/pdf/conveng.pdf.

Vidal, John. 2009. 'Ed Miliband: China tried to hijack Copenhagen climate deal'. *The Guardian*, 20 December. Accessed 26 February 2018. www.theguardian.com/environment/2009/dec/20/ed-miliband-china-copenhagen-summit.

Vihma, Antto, Yacob Mulugetta & Sylvia Karlsson-Vinkhuyzen. 2011. 'Negotiating solidarity? The G77 through the prism of climate change negotiations'. *Global Change, Peace & Security* 23:3, 315–334.

Voigt, Christina. 2008. 'State responsibility for climate change damages'. *Nordic Journal of International Law* 77, 1–22.

Wang, Yi. 2013. 'Exploring the path of major-country diplomacy with Chinese characteristics'. Ministry of Foreign Affairs, People's Republic of China, 27 June. Accessed 29 May 2017. www.fmprc.gov.cn/mfa_eng/wjb_663304/wjbz_663308/2461_663310/t1053908.shtml.

Wara, Michael. 2007. 'Is the global carbon market working?'. *Nature*445, 595–596.

White House. 2017. 'Statement by President Trump on the Paris climate accord'. Accessed 25 September 2017. www.whitehouse.gov/the-press-office/2017/06/01/statement-president-trump-paris-climate-accord.

White House. 2014. 'U.S.-China joint announcement on climate change'. On file with author.

White House. 2013. 'Joint U.S.-China statement on climate change'. On file with author.

Wong, Sue-Lin. 2016. 'In rare move, China criticizes Trump plan to exit climate change pact'. Reuters, 1 November. Accessed 22 February 2018. www.reuters.com/article/us-climatechange-china/in-rare-move-china-criticizes-trump-plan-to-exit-climate-change-pact-idUSKBN12W349.

World Economic Forum. 2017. 'President Xi's speech to Davos in full'. Accessed 6 February 2017. www.weforum.org/agenda/2017/01/full-text-of-xi-jinping-keynote-at-the-world-economic-forum.

World Meteorological Organization. 2017. *WHO Greenhouse Gas Bulletin*, no 13. Accessed 6 February 2018. https://public.wmo.int/en/resources/library/wmo-greenhouse-gas-bulletin.

Yamin, Farhana & Joanna Depledge. 2004. *The International Climate Change Regime: A Guide to Rules, Institutions and Procedures.* Cambridge: Cambridge University Press.

Xi, Jinping. 2017a. 'Work together to build a community of shared future for mankind'. Xinhua News Agency, 19 January. Accessed 6 February 2017. http://news.xinhuanet.com/english/2017-01/19/c_135994707.htm?from=singlemessage.

Xi, Jinping. 2017b. 'Secure a decisive victory in building a moderately prosperous society in all respects and strive for the great success of socialism with Chinese characteristics for a new era'. Xinhua News Agency, 18 October. Accessed 14 February 2018. www.xinhuanet.com/english/download/Xi_Jinping's_report_at_19th_CPC_National_Congress.pdf.

Xi, Jinping. 2017c. 'Work together to build the Silk Road economic belt and the 21st century maritime Silk Road'. Xinhua News Agency, 16 May. www.xinhuanet.com/english/2017-05/16/c_136287878.htm.

Xi, Jinping. 2015. 'Work together to build a win-win, equitable and balanced governance mechanism on climate change'. Ministry of Foreign Affairs, People's Republic of China Accessed 14 February 2018. www.fmprc.gov.cn/mfa_eng/wjdt_665385/zyjh_665391/t1321560.shtml.

Xi, Jinping. 2014. 'Seek sustained development and fulfill the Asia-Pacific dream'. Ministry of Foreign Affairs, People's Republic of China. Accessed 14 February 2018. www.fmprc.gov.cn/mfa_eng/topics_665678/ytjhzzdrsrcldrfzshyjxghd/t1210456.shtml.

Zhang, Gaoli. 2014. 'Build consensus and implement actions for a cooperative and win-win global climate governance system'. Ministry of Foreign Affairs, People's Republic of China Accessed 14 February 2018. www.fmprc.gov.cn/mfa_eng/zxxx_662805/t1194637.shtml.

6 The fulfilment of China's climate responsibility

Responsibility must be demonstrated by action. It is not enough to discuss climate responsibility; on the contrary, such responsibility must be acted upon by formulating and implementing ambitious climate policies at the domestic level. In this chapter, I investigate the ways in which China plans to fulfil it climate responsibility in practice. Because climate policy cannot be viewed as somehow unrelated to states' overall policies and goals, I also elaborate upon how the Chinese government's grand strategy to develop China as a strong, prosperous and modern socialist country shapes its domestic climate policies. I begin by examining drivers of China's climate policy because such factors indicate the extent of the Chinese government's commitment to take action against climate change as well as to adapt to such change. Later, I review China's international climate commitments under the United Nations Framework Convention on Climate Change (UNFCCC) and the ways by which China has proposed to uphold those commitments by implementing, for instance, low-carbon development plans and energy policies.

Drivers of China's climate policy

As discussed in chapter 2, three drivers motivate states to participate in international practices: coercion, calculations and identity and belief (Buzan 2004, 103, 130–133, 253–261; Hurd 1999; Hurd 2007, 30–40; Hurrell 2007, 67–77; Wendt 1999, 247–250). The ways in which states adopt new norms and practices influence how profoundly, if at all, states incorporate them into their identities. In the case of coercion, norms are only superficially internalised or incorporated into a state's identity, and the state is therefore unlikely to assume the responsibilities required by the corresponding practices. Calculations of self-interest cannot sufficiently motivate the fulfilment of responsibilities either; they are unlikely to cause profound changes in an actor's preferences and values, and social practices relying on self-interest necessarily have weak foundations. Conversely, if states that engage in an international practice believe that the practice is legitimate, then they become willing to shoulder their fair share of responsibility to implement the norm. The perceived legitimacy of norms and practices is thus an essential driver of assuming responsibility. In reality, however, that typology can seem impractical; after all, all three reasons motivate human conduct and are often difficult to differentiate. As Buzan (2004, 130) concludes, the combination of the three reasons therefore makes politics what it is. Accordingly, in this section I elaborate upon the underlying motives of China's climate policies and ask to what extent those policies respond to external pressure, calculations of self-interest or identity and belief. Ultimately, the results can clarify what sort of responsibility China will be willing and able to shoulder in international climate politics, as well as how it will enact that responsibility, if at all.

Coercion and regulation

As a driver of state practices, coercion does not play an important role in international climate politics. Participation in international climate negotiations is voluntary, and international treaties on climate change do not have legal status strong enough to coerce states to take certain actions to mitigate climate change. When entering into any international treaty, including one concerning climate change, a state is expected to comply with the treaty's regulations, although no sanctions follow if it fails to meet those expectations. For China, international climate treaties do not impose obligations that could somehow be viewed as having coercive force on its conduct. When the Kyoto Protocol set quantitative emissions reduction targets for developed countries only, China had no obligation to cut emissions. Moreover, because the Paris Agreement is based on states' voluntary commitments to reduce emissions, China has thus been able to independently set its own national targets. Although social pressure can be somewhat characterised as playing a coercive role in inter-state relations, international criticism of China's climate irresponsibility has not forced China's government to alter its attitude towards climate politics. However, its calculations of losses to China's image have played an important role in the state's response to such criticism.

Calculations

Calculations of self-interest stand as clear drivers of China's climate policy, as they do for all other states. As discussed in previous chapters, national interests are not static or fixed but shift as circumstances change. For centuries, maintaining stability and unity has been a core interest of China's leaders, and at present, supporting economic growth remains central to the maintenance of China's domestic stability. Therefore, the Chinese government stresses its moral responsibility to maximise economic growth, which constitutes its chief priority in all policies because China has to 'complete the historical task of industrialization and urbanization' (Information Office of the State Council of the People's Republic of China 2011). Economic growth is thus the major driver of China's conduct in every sector of political life. At the same time, China's vast economic growth has caused severe environmental problems, including water and air pollution, deforestation, desertification, hazardous waste and losses of biodiversity.[1] Since many Chinese citizens have become ill or even died due to air, land and water pollution, environmental problems represent a top reason for social discontent in contemporary China. Educated middle-class citizens in particular have increasingly urged the Chinese government to resolve problems caused by smog – so-called 'airpocalypses'– due to industrial pollution. To maintain social stability and thus its own legitimacy, the party-state has consequently had no choice but improve standards of environmental protection. Since the burning of coal is a chief factor of air pollution in China, energy policy plays an important role in those efforts.

As norms of climate responsibility begun to evolve, China played a conspicuously marginal role in international relations pertaining to greenhouse gas emissions and international status. Between the Stockholm and Rio conferences, however, China gained substantial leverage in its role in international environmental politics. The world began to recognise China's critical role in global environmental change, and although China continued to be perceived as a poor country, its large size and population were identified to bear significant influence over environmental issues that, once purely local, had become global problems requiring urgent international cooperation. At the same time, as China realised its environmental power, it began to regularly participate in international environmental negotiations. The Chinese government quickly acknowledged China's interest in participating in the formulation of international environmental practices, which it viewed as an outstanding opportunity to attract foreign aid and spur technology transfer (Kobayashi 2005, 95–96), as well as equip Chinese officials and scientists with valuable training from international environmental organisations that could aid China's capacity to solve domestic environmental problems (Economy 1998, 274). From another angle, the government considered it more beneficial to participate in developing rules for international environmental practices than to merely comply with rules and norms set by others. In its calculations of self-interest, China's government perceived an opportunity to expel Taiwan from international organisations and thereby defend the status of the People's Republic of China as China's legitimate representative at the international level. Moreover, because the Tiananmen Square incident on 4 June 1989 had disastrously affected China's image in international politics, the Chinese government regarded environmental issues as being 'soft' enough to restore the state's international image and its capacity to rejoin international society (Harrington 2005, 110). Indeed, soon after the incident, China showed its willingness to engage in international environmental cooperation when more than 120 Chinese environmental groups visited their counterparts abroad and when, in 1990 alone, foreign delegations from more than 50 nations were received to China (Shouqiu & Voigts 1993, 24). Stemming from a fear of international isolation following the incident, China also expanded its cooperation with developing countries (Kobayashi 2005, 96).

Despite China's burgeoning economic prosperity and enhanced international status, the principle of common but differentiated responsibilities, climate finance and technology transfer continue to occupy central positions in its agenda regarding international climate negotiations. Although China took an obvious position of great power at the UN Climate Change Conference in Paris in 2015, it also sought to re-establish a rigid division between developed and developing countries during negotiations over the Paris Rulebook. Clearly, China is unlikely to sign an international treaty that would hinder its industrialisation and thus impede its development by, for example, setting per capita emission levels that are too low. China shows a justifiable fear that emissions controls could increase business costs in China and that investment and jobs important to its economic growth would thus be relocated to countries where labour is cheaper. Accordingly, China aligns its interests with the interests of all developing countries; because developed countries cannot deny the assistance-related needs of developing countries as a group, it is rhetorically more convincing to articulate the needs of that group instead of China's own requirements. If developed countries admit the major role of funding and technology transfer in mitigating climate change in developing countries, China, still categorised as a developing country, receives assistance and Clean Development Mechanism credits.

By casting itself as the defender of the developing world's interests, China has boosted its image among the political leaders of developing countries. To further differentiate China from Western countries, the Chinese government often emphasises China's common history with developing countries as victims of Western colonialism and exploitation. In recent years, China has shown substantial goodwill to developing countries and increased its soft power there by offering loans and grants, debt relief, weapon sales, student scholarships and assistance with infrastructure projects that build roads, schools and housing. Such efforts are important to China's pursuit of comprehensive power, because developing countries represent a considerable source of potential diplomatic support for China's status in international politics, especially in international organisations that maintain a one country, one vote system. They could therefore provide China with important diplomatic support in its pursuit to democratise international relations, increase its leverage in international politics and oppose Western standards of civilization and other Eurocentric practices of the US-led world order. Moreover, China is the only developing country with a permanent seat on the UN Security Council, which enables it to mentor other developing countries in international affairs. In contrast to US soft power, Chinese soft power is interactive and 'derives from Beijing's courtship and what regional neighbors perceive as mutual benefits' (Percival 2007, 111–113). Because China's soft power is often dismissed at the international level – Chinese political ideals and policies do not typically uphold US ideals and values such as human rights and democracy – an important source of such power is the state's successful development model and new breed of diplomacy. In that sense, because many developing countries feel marginalised by the current US-led international society, they view China as a generator of economic and political opportunity. In particular, many socialist and authoritarian regimes perceive the Chinese model as an especially attractive alternative to democratisation. At the same time, China's 'friendship with various dictators' decreases its status and prestige among other international powers (Ding 2008, 201–202).

Clearly, China's climate policy is also partly motivated by international pressure. As Alastair Iain Johnston (1998, 559) explains, 'the more international criticism China meets or is likely to meet on some issue of international cooperation, the more likely it will try to find ways to minimise this criticism through incrementally substantive cooperative commitments'. Such actions are largely motivated by the pursuit of a favourable international image, or face, which has long guided Chinese foreign policy (e.g. Brady 2012; Deng 2008; Kopra 2012; Wang 2005).[2] China's government not only wants to assure the world that China poses no threat and that other states will even benefit from its development but also seeks to gain face among other states, which can support two values central to the domestic legitimacy of the Communist Party of China (CCP): authority and national honour. Although China did not significantly contribute to many environmental problems before the late 1990s, it nevertheless attended the UN Conference on Environment and Development (UNCED) and the UN Conference on the Human Environment, arguably to nurture its international image (Economy 1998, 269). After the 2009 UN Climate Change Conference in Copenhagen, at which China encountered harsh criticism for obstructing international cooperation, constructing a favourable national image resumes its influence over the state's climate policy. Most recently, at the Paris Conference in 2015, China's demonstrated its preference for the moderate, voluntary commitments typically pursued during the meeting's negotiations over legal international obligations, for the former not only dispel China's fear of failure but also allow it to easily exceed global expectations and thus gain face. Of course, national image building is nothing new in Chinese foreign policy. China's Century of Humiliation (1839–1949) and its loss of status as the most sophisticated civilisation in the world have been central to its national identity and important reasons for its eagerness to restore its status internationally. Nevertheless, as Johnston (1998, 560) argues, it is difficult to pinpoint 'where exactly this concern about image or reputation comes from, or how image is to be turned into a cost or benefit'. From the perspective of the English School, one benefit of having a reputation as a responsible nation would be acceptance as a member of the great power club, which would award China greater rights on the international stage.

Although China's attention to national image building has precedents, its confidence in its emerging role as a leader in international politics is altogether new. China no longer responds only to international expectations; since US President Donald Trump's decision to withdraw from the Paris Agreement, China has proactively portrayed itself as the global leader in climate politics. Presumably, one reason for the shift is the approaching centennial of the CCP. As discussed earlier, the first of China's centenary goals expects the state to be a strong international player in all sectors of international society by 2020. In many fields of foreign policy, including engagement in Africa and the Arctic, both of which have been viewed as the property and backyard of the West for historical and geopolitical reasons, China's growing role has raised international concerns, and it has been difficult for the party-state to assume the role of leader. By contrast, mitigating climate change does not hamper China's national interests but instead offers plenty of economic and imagological benefits for the party-state. Among the doors that President Trump's America First approach has opened for China's emerging global leadership, a position at the fore of climate politics allows the party-state to represent itself as a strong global leader in the eyes of domestic audiences. Although Trump's policy has thus unintentionally helped China's leaders to achieve their centenary goals, China's emerging role as leader per se is not accompanied by new ambition in international climate politics. Because China will not necessarily be an inspirational leader that will or even can advance ambitious action against climate change worldwide – after all, leaders can just as well advance harmful policies and practices that do not promote human values – what sort of global leader China will become thus depends on China's identity.

Identity and belief

Belief has driven China's climate policy since the early 1990s. In general, China's political leaders do not seem to dispute whether climate change is real or whether it is caused by human action. Published in 1990, the first report of the Intergovernmental Panel on Climate Change generated political will in China to participate in international efforts to prevent climate change. Although officials in the Ministry of Foreign Affairs emphasised China's status as a developing country and thus underscored the responsibilities of developed countries to lead efforts to allay climate change, many Chinese citizens believed that China, as a member of international society and 'primarily because the PRC, too, would be affected', should contribute to international climate politics (Economy 1998, 271). Although that stance was largely based on calculations of self-interest, it nevertheless indicated the increasing belief among the Chinese that climate change poses a serious risk both locally and globally. Indeed, the Chinese expressed a largely solidarist perspective, for they believed that as a member of international society, China had a responsibility to act with regard to global environmental problems (ibid.). Accordingly, when China decided to actively participate in climate negotiations as a leading representative of developing countries, the state's official position in the negotiations incorporated a synthesis of all three perspectives. The official position, however, more or less ignored the most solidarist perspectives, including the one expressed by Song Jian, then chair of the State Science and Technology Commission, in 1991 (quoted in Economy 1998, 276):

> As we [the Chinese] develop the economy, we must guarantee a balanced ecological environment and maintain in good order our natural resources so that future generations will have their rightful heritage. To this end, we should be ready to pay more or, if necessary, slow down the economic development.

For the Chinese government, climate change is an 'issue involving both environment and development, but it is ultimately an issue of development' (National Development and Reform Commission 2007). Underlying that characterisation are two beliefs: that climate change has been caused by the historical development processes of developed countries and that it prevents developing countries from achieving prosperity. Accordingly, China's government has maintained that the 'ultimate solution to climate change can only be achieved through common sustainable development of all countries' (Xie 2010). In effect, China's stance on climate change influences the state's position in international climate politics in two ways. On the one hand, China expects developed countries to transform their consumption habits and implement ambitious emissions cuts; on the other, it stresses the ability of developing countries to adapt, albeit with the assistance of developed countries, to climate change (Kopra 2016). At the same time, China's approach to climate change is exceptionally technocratic; the government has even argued that 'technology innovation and transfer are the basis and support for addressing climate change' (Information Office of the State Council of the People's Republic of China 2008). Altogether, China's approach to climate change suggests that global responsibility for mitigating such change depends upon a country's developing or developed status, not the global impacts of its policies and actions – in other words, that responsibility should be allocated according to emissions per gross domestic product (GDP).

China's government moreover believes that climate change is a 'challenge faced by the entire world' and thus can be solved only by 'extensive international cooperation' (Information Office of the State Council of the People's Republic of China 2008). Therefore, it has engaged in UNFCCC negotiations since they began in the late 1980s and regards the UNFCCC and its Kyoto Protocol as the 'most authoritative, universal and comprehensive international framework for coping with climate change' (ibid.). The Chinese government has also recently begun to recognise the security-related impacts of climate change and believes that China ranks among 'the countries most vulnerable to the adverse impact of climate change' (National Development and Reform Commission 2012). Although it does not regard climate change as an issue of national security, the Chinese government is aware that it harms China's economic and social development, as well as the lives of its citizens (National Development and Reform Commission 2014, 4). In 2011, for instance, extreme weather events and natural disasters related to climate change affected 430 million Chinese people and caused 309.6 billion yuan in economic losses (National Development and Reform Commission 2012). Moreover, Nicholas Stern (2006, 106) estimates that, from 1988 to 2004, 'China experienced economic losses from drought and flood equating to 1.2 per cent and 0.8 per cent of GDP respectively'. On top of such losses, climate change also seriously jeopardises food security in China.

At present, China's national identity is in flux. Both the Chinese general public and its leaders believe that 'China is a nation with a dual-identity' – both a developing country and a major power (Wu 2001, 293) – and that belief has significantly complicated domestic consensus on the scope of China's global responsibility. Although many Chinese citizens consider that the state should assume more international responsibility, many others believe that it is unjustifiable to expect China to shoulder more global responsibility before it has achieved a greater stage of development. China's role in international climate politics reflects that dualism; not only have the Chinese tended to discursively construct a highly dualist image and identity for the state (Kopra 2012), but China has also participated in various, even conflicting, coalitions in climate politics. As discussed earlier, China has consequently begun to describe itself as a great power in international climate politics and assumed a constructive role in international climate negotiations since the 2009 Copenhagen Conference. During the Obama administration, China and the United States actively cooperated to develop policy against climate change, which significantly influenced China's attitudes towards climate responsibility, particularly the Chinese government's view that great powers have great responsibility to mitigate climate change. However, when cooperation on climate policy between the great powers ended during the Trump administration, China began to position itself as the defender of developing countries' interests at the 2017 UN Climate Conference, where it sought to resume negotiation about the bifurcation of developed and developing countries. China thus confirmed its role as the leader of the developing world, a role which has 'sometimes forced it to be more uncooperative in international environmental affairs that it would actually like to be' (Kobayashi 2005, 88). In particular, China's participation in so-called 'spoiler' clubs in international politics, including the Like-Minded Countries and the BASIC (Brazil, South Africa, India and China), prevents the state from taking more responsibility in international society. When US–Chinese cooperation on mitigating climate change dissolved, China and the European Union intensified their cooperation to the same end, and time will reveal whether such collaboration encourages China to emphasise its status as a great power in international negotiations on climate change. For now, it is clear that identity politics matter enormously in international climate politics.

China's commitment to mitigate climate change

Until 1998, the China Meteorological Administration was responsible for advising the Chinese government about issues in the international arena related to climate change. Ever since, such responsibility has fallen to the powerful State Development and Planning Commission (renamed the National Development and Reform Commission in 2003), which has indicated China's shift from viewing climate change as a scientific issue to viewing it as a political and economic problem (Heggelund, Andresen & Fritzen Buan 2010, 237–238). In 2008, China upgraded the State Environmental Protection Administration to a full-fledged Ministry of Environmental Protection not only responsible for organising national environmental policies and preventing pollution but also with a mission to 'shoulder and materialize the responsibility for achieving national target on emission reduction' (Ministry of Environmental Protection 2008). Today, although both the Ministry of Environmental Protection and the Ministry of Science and Technology participate in China's climate-focused diplomacy, the 'hardliner' Ministry of Foreign Affairs plays a central role in China's international climate politics (Heggelund, Andresen & Fritzen Buan 2010, 239).

Until recently, China had refused to make any commitments in international negotiations addressing the climate. In 2006, when it overtook the United States in total greenhouse gas emissions and became the largest emitter of carbon in the world, its leverage in such negotiations afforded it, along with the United States, a 'position to make or break' international climate agreements (Harris 2013, 76). Since the Copenhagen Conference in 2009, China has indeed taken a more constructive role in international climate negotiations. Although the Chinese government stated in 2010 that China 'could not and should not' set a carbon cap (Lan 2010), it had committed to pursue actions against climate change proportionate to its size and population at the Bali Conference in 2007. According to China's nationally appropriate mitigation actions commitment,

> China will endeavor to lower its carbon dioxide emissions per unit of GDP by 40–45% by 2020 compared to the 2005 level, increase the share of non-fossil fuels in primary energy consumption to around 15% by 2020 and increase forest coverage by 40 million hectares and forest stock volume by 1.3 billion cubic meters by 2020 from the 2005 levels.
>
> (Su 2010)

By the 2015 Paris Conference (COP21), China's position in international climate negotiations had transformed. Prior to COP21, China had announced important commitments to reduce its greenhouse gas emissions, and in November 2014, it and the United States signed a historic agreement in which it committed to stem the growth of its CO_2 emissions by 2030 (White House 2014). As a result, China no longer focuses on reducing relative carbon intensity (i.e. amount of CO_2 per unit of GDP) but has instead pledged to reduce its absolute emissions. Such resolution sent a strong signal to international society that both the United States and China had acknowledged their responsibility to lead international efforts towards tackling climate change and that an international climate treaty was possible. In June 2015, China published its Nationally Determined Contribution (NDC) to the UNFCCC, in which it reiterated its commitment to not exceed CO_2 emissions achieved by 2030, to reduce its carbon intensity by 60–65 per cent of its 2005 level by 2030, increase the share of non-fossil fuels used in primary energy consumption to roughly 20 per cent and boost the forest stock volume by approximately 4.5 billion m^3 above the 2005 level (National Development and Reform Commission 2015, 5). China's NDC, however, neither sets a cap for emissions nor specifies how much Chinese emissions will climb before they peak. China also has not set any long-term goal to explain how much and by when the state expects to reduce greenhouse gas emissions after 2030. Instead, it stresses the principle of common but differentiated responsibilities and reminds developed countries of their historically informed responsibility to 'undertake ambitious economy wide absolute quantified emissions reduction targets by 2030' and 'provide support for developing countries to formulate and implement national adaptation plans as well as other related projects' (ibid., 17–18).

China's government assures other states that China's NDC is motivated not only by national interests but also by its 'sense of responsibility to fully engage in global governance, to forge a community of shared destiny for humankind and to promote common development for all human beings' (National Development and Reform Commission 2015, 2). In any case, China will benefit from fulfilling its NDC by reducing its reliance on coal to generate power, decrease air pollution and thereby prevent 100,000 premature deaths annually and create an additional half-million full-time green jobs in its renewable energy sector (Höhne et al. 2015, 21). However, China's NDC does not explain why or how it is fair and ambitious, which is necessary to inspire similarly ambitious climate policies of other states, including developed ones that often cite China's climate irresponsibility for their own inaction. A day after the publication of China's NDC, China's National Center for Climate Change Strategy and International Cooperation published a commentary demonstrating that China has nevertheless considered the ethical dimensions of its NDC. At the same time, the commentary not only describes the ambitiousness and economic, technological and social challenges of climate policies but also again highlights China's status as a developing country. It therefore suggests that if the state manages to curb emissions at a faster rate after 2030, China's long-term development path will align with the objective of limiting global average temperature increase to less than 2°C (Fu, Ji & Liu 2015, 7–8). However, the Climate Action Tracker (2018), an independent scientific analysis team representing four research organisations, argues that China's NDC will not limit global warming to less than 2°C unless other states commit to much more ambitious emissions reductions than China's. Therefore, the Climate Action Tracker (2018) gives China a rating of 'highly insufficient'.

China has not joined the Green Climate Fund, and the 2015 Paris Conference did not oblige China, as a developing country, to contribute financial assistance, although it encouraged 'other parties' to 'provide or continue to provide such support voluntarily' (UNFCCC 2015). Prior to the conference, however, China had pledged to establish its own voluntary and complementary South–South Cooperation Fund, which will 'make available ¥20 billion [about 3.1 billion USD] for setting up the China South–South Climate Cooperation Fund to support other developing countries to combat climate change' (White House 2015a).

China's domestic climate policies

The UNFCCC (1992) does not specify how states should fulfil their climate responsibility but asks them to take action appropriate to their national circumstances. In practice, each state therefore decides what sort of policies it will undertake in order to meet its climate responsibility; a state may choose any combination of market mechanisms (e.g. carbon trade and taxation), technological solutions (e.g. carbon capture and storage), voluntary and mandatory emissions limits and education. For its part, China has primarily chosen traditional tools such as legislation and regulation, central government planning and government-led projects and programmes to uphold its climate responsibility at the domestic level. Because the National Development and Reform Commission sets China's domestic agenda on economic, energy and climate policies, China's climate policy has unsurprisingly prioritised strategies amenable to economic development and energy use and stressed the principle of development first.

Low-carbon development

Although the concept of sustainable development was a major issue of debate at UNCED, no consensus was reached regarding how such development should be achieved in practice (cf. Hopwood, Mellor & O'Brien 2005). After UNCED, China embraced the concept but has seemed to focus on its economic aspects. For example, in 1996, China's ninth Five-Year Plan confirmed sustainable development as a national development strategy but focused on economically sustainable development while environmental protection was viewed merely as a tool to achieve economic development. According to then Chinese premier Li Peng (1995), China should 'rationally develop and utilize resources and protect the ecological environment so as to achieve a coordinated and sustainable economic and social development'. Accordingly, in one of the first such state documents worldwide written in accordance with UNCED, China's Agenda 21 proposed a comprehensive approach to environmental protection by integrating policies such as economic development, industrialisation, population control, agriculture, energy production, education, health and sanitation, disaster management and the protection of the atmosphere (Economy 1998, 276). In effect, China's Agenda 21 demonstrates not only that the Chinese state did not dispute whether climate change is real even in the early stages of climate protection but also that the Chinese government took the issue of climate change seriously.

In the early 2000s, China began to examine how it could modernise in a more sustainable way and thereby alleviate social and environmental problems caused by its development model (Dent 2014, 57). In particular, the concepts of scientific development and the so-called 'harmonious society' have served as ideological guidelines for the state's development policies (cf. Fan 2006, 709). For instance, China's 11th Five-Year Plan (2006–2010), in projecting 'big change in the relationship between the environment and development' in China, announced 'three transformations' that would encapsulate China's commitment to balance economic growth with environmental protection. The first two transformations downgrade the previous priority of economic development to place equal emphasis on both it and environmental protection. The third transformation, which China arguably experiences at present, results from 'mainly employing administrative methods to protect the environment into comprehensive application of legal, economic, technical and necessary administrative methods to address environmental problems' (Ministry of Environmental Protection 2008). The 11th Five-Year Plan also states that 'economic growth is not the equivalent of economic development' (Fan 2006, 710) but that environmental protection is a 'key task for modern development' (Ministry of Environmental Protection 2008). Since then, China has paid special attention to the 'greening of the economy' and the development of 'green jobs' (Pan, Ma & Zhang 2011).

Beginning in the late 2000s, the Chinese government commenced implementing several policies to reduce the country's greenhouse gas emissions. In June 2007, the government published its first comprehensive climate policy, the National Climate Change Programme, which pledged that 'China will implement its fundamental national policy of resources conservation and environmental protection to develop a circular economy, protect the ecological environment and accelerate the construction of a resource-conservative and environmentally-friendly society' (National Development and Reform Commission 2007). The policy also promised to 'strive to control its [China's] greenhouse gas emissions, enhance its capacity to adapt to climate change and promote the harmonious development between economy, population, resources and the environment' (ibid.). In his report to the 17th Party Congress, President Hu Jintao (2007) promised that China would 'make new contributions to protecting the global climate' and 'implement the responsibility system for conserving energy and reducing emissions'. Consequently, the government issued the first white paper on climate change in 2008, and in August 2009, the National People's Congress of China Standing Committee passed a resolution to actively respond to climate change, which marked the first climate change resolution adopted by China's top legislature body. In particular, the resolution stressed the principle of scientific development and vowed to strengthen China's legal framework for climate change by declaring that

> Responding to climate change is an important opportunity and challenge for China's economic and social development. To actively respond to climate change is of great importance to China's overall economic and social development, the people's vital interests, human survival and the development of all countries.
>
> (Standing Committee of the National People's Congress 2009, my translation)

In November 2009, to prevent greenhouse gas emissions from doubling by 2020, China announced a voluntary but nationally binding target to reduce carbon intensity by 40–45 per cent of 2005 levels by 2020 (Xinhua 2009). In March 2011, the target was incorporated into the 12th Five-Year Plan (2011–2015), which also not only determined to lower energy consumption per unit of GDP by 16 per cent by 2015 and CO_2 emissions by 17 per cent but moreover increase the proportion of non-fossil fuels in overall primary energy consumption to 11.4 per cent compared to 2010's 8.3 per cent. Perhaps most dramatically, in 2014 Chinese Premier Li Keqiang announced that China had declared war on pollution and would first focus on reducing the levels of $PM_{2.5}$ and PM_{10} in the atmosphere by improving energy efficiency, raising the proportion of electricity generated by renewables and nuclear power, developing low-carbon technology, reducing vehicle exhaust emissions and shutting down outdated industrial plants and energy producers (Li 2014). Doubtless, all of those efforts were expected to contribute to mitigating climate change as well.

Later in 2014, the State Council approved China's National Climate Change Plan (2014–2020), which introduced various measures to stem the growth of greenhouse gas emissions, promote low-carbon development plans, improve the state's adaptation to climate change and expand international cooperation in mitigating climate change. In particular, the plan prioritised efforts to reduce greenhouse gas emissions in both the industrial sector and the construction industry. Although it did not set a binding, nationwide emissions reduction target, the plan did seek to reduce CO_2 emissions per unit of industrial added value by 50 per cent below 2005 levels by 2020. Notably, the plan also pledged to stabilise the total carbon emissions of the steel and cement sectors at 2015 levels by 2020 (National Development and Reform Commission 2014, 10–11). Although the plan's various targets sent a positive signal to the world about the government's strong commitment to mitigating climate change, the plan's feasibility received criticism due to the lack of coordination and motivation among government sectors outside the NRDC

(Liu & ClimateWire 2014). If met, however, the plan's targets would halt China's rapid emissions growth. In the construction industry especially, the plan promotes low carbon technologies and materials and aims for half of newly constructed urban buildings to be green by 2020 (National Development and Reform Commission 2014, 13). Other efforts to reduce China's greenhouse gas emissions include promoting the use of renewable energy sources, improving energy efficiency, controlling emissions in transportation and agriculture, increasing carbon sinks and promoting low-carbon lifestyles.[3]

Since the 18th National Congress in 2012, China has actively pursued the development of what it calls 'an ecological civilisation', which China's climate policies have also listed as an important objective (National Development and Reform Commission 2014, 3). Published in 2015, 'An Opinion on Acceleration for the Promotion of Ecological Civilization' not only articulates objectives and instructions for reorganising the Chinese economy to make it greener but also addresses ways to overcome obstacles to improving the environmental standards of Chinese society. In its 12,000 characters, the opinion even declares that 'green development' (绿色发展) is important to the development of China's competitive advantage and thus plays an important role in fostering comprehensive national power. That same year, given the State Council's aim to make 'significant progress' in cultivating a resource-saving, ecofriendly society by 2020, China's government relaunched the green GDP project and announced plans to start pilot projects at the regional level in the near future. The State Council maintains that CCP committees and governmental officials at every level are responsible for constructing China's ecological civilisation. In particular, it highlights that economic performance should no longer be the 'only criterion in government performance assessment' but that environmental issues should have greater weight in cadre evaluations. At a more personal level, the State Council added that cadres have a lifelong responsibility (终身追责) for environmental harm caused on their watch, even after they have retired or changed positions. Accordingly, if they manage to promote the construction of an ecological civilisation, then they will be rewarded; however, if they fail or cause serious environmental damage, then they cannot be promoted to higher positions (Xinhua 2015a).

In September 2015, the State Council and the Central Committee, the highest bodies in the Chinese government and the CCP, in jointly publishing the 'Integrated Reform Plan for Promoting Ecological Progress' indicated the government's strong political will to improve environmental protection in China.[4] The plan not only insists that ecological conservation be given a 'position of prominence and incorporated into every aspect and the whole process of economic, political, cultural, and social development' but also presents a comprehensive structural framework of institutional reforms needed to improve the 'formation of a new pattern of modernization in which humankind develops in harmony with nature' (Xinhua 2015b). Although the plan confirms that it 'is necessary to remain committed to the strategy of treating development as being of the utmost importance to China', it nevertheless acknowledges that development 'is good only when it is green, circular, and low-carbon' (ibid.). Moreover, it reiterates the lifelong responsibility of cadres to environmental conservation and proposes the option to develop a national supervision and inspection system to perform their 'natural resource asset audits' (ibid.). To mitigate climate change, the plan announces ambitious objectives to be achieved by 2020. First, it pledges that a 'system for controlling total national carbon emissions and a mechanism for breaking down the responsibility for implementation will be gradually established', which can be interpreted as the Chinese government's commitment to adopting an absolute emissions reduction target after 2020. Second, it declares that 'subsidies for all fossil fuels will be phased out' (ibid.); such a measure sets a tremendous example for other states, which currently spend roughly 5 trillion USD annually to subsidise the consumption of fossil fuels, an amount that is 'over four times the value of subsidies to renewable energy and more than four times the amount invested globally in improving energy efficiency' (International Energy Agency 2014). Last, the plan promises to invest more in the development of renewable energy sources and establish a 'mechanism for effectively increasing forest, grassland, wetland, and ocean carbon sinks' (Xinhua 2015b).

For the time being, China's 13th Five-Year Plan (2016–2020) forms the most significant strategic guidelines for the state, including those pertaining to practices of climate responsibility. Above all, the current plan integrates China's international climate commitments into the overall development objectives and measures of the party-state. In particular, it seeks to reduce China's carbon intensity by 18 per cent from 2015 levels and reiterates the national energy consumption cap of 5 billion metric tonnes of standard coal equivalent by 2020. Notably, it proposes to not only implement but also *enhance* the state's nationally determined contribution to the UNFCCC (National People's Congress & Chinese People's Political Consultative Conference 2016). At present, some Chinese cities and provinces have also outlined exceptionally ambitious plans to respond to climate change. At the US–China Climate Leaders Summit held in Los Angeles in September 2015, 11 Chinese cities and provinces committed to reaching their peak greenhouse gas emissions before the national target of 2030 and resolved to establish the corresponding Alliance of Peaking Pioneer Cities. In particular, two of the most populous cities, Beijing and Guangzhou, pledged to reach their peak CO_2 emissions by or around 2020 (US–China Climate Leaders' Declaration 2015). As the fact sheet of the White House (2015b) notes, 'The commitment of so many of its [China's] largest cities to early peaking highlights China's resolve to take comprehensive action across all levels of government to achieve its national target, put forth in last year's Joint Announcement on Climate Change'.

Energy policy and beyond

The demand for energy in China is massive because the country's economic growth is largely based on fuel-intensive heavy industry. Given the risks of energy shortages, energy security has been a top priority for China since the late 1990s. At present, with coal as 60 per cent of its energy mix, China uses more coal than any other nation in the world. Because such use causes not only enormous carbon emissions but also severe air pollution around the country, both domestic pressures and international commitments motivate the Chinese government to decrease the share of coal in the country's energy mix and to develop alternative energy sources. In response, the government has issued a series of policies and measures to decrease China's dependency on coal and other (imported) fossil fuels, as well as to promote the production of non-fossil energy, especially hydropower and nuclear energy. The government has also made serious efforts to decrease energy demands by promoting energy conservation and energy efficiency. To that end, it has closed ineffective power plants and small or outdated industrial factories, promoted the development of modern, energy-saving technology and products and established national standards to improve automotive fuel economy.

For the first time, China's current Energy Development Strategy Action Plan (2014–2020) includes a cap on national coal consumption by 2020 and pledges to raise the share of non-fossil fuels in the total primary energy mix to 15 per cent by 2020 from 2013's 9.8 per cent (Xinhua 2014). Likewise, the 13th Five-Year Plan orders the increased proportion of non-fossil energy to 15 per cent and the decreased consumption of coal to less than 55 per cent. In keeping with those goals, the government has also decided to abandon or delay at least 150 gigawatts of coal-fired power projects by 2020 (Reuters 2016). At the same time, China doubtlessly already stands as the world leader in renewable energy, including solar and wind power, hydropower and bioenergy for electricity and heat, as well as a pioneer in the development of electric vehicles. In fact, the International Energy Agency (2017) estimates that China will account for 40 per cent of global capacity growth in renewable energies between 2017 and 2022.

In terms of hydropower, China has the richest resources in the world and has efficiently taken advantage of them. In particular, China's installed capacity of hydropower exceeded 100,000 megawatts in 2004, 200,000 megawatts in 2010 and 300,000 megawatts in 2015 (Li et al. 2018). Although China is the greatest source of hydropower worldwide, its massive hydropower projects have also caused severe social and environmental harms at the local level (e.g. Zhao et al. 2012). For example, the Three Gorges Dam is famous not only for its massive size but also because of the vast social and environmental problems it has caused. By contrast, the development of small hydropower resources (i.e. hydropower stations with an installed capacity of no more than 50 megawatts) seems to pose outstanding potential to resolve energy and environmental problems as well as alleviate poverty in rural China (e.g. Kong et al. 2015; Kong et al. 2016).

In terms of solar power, China is also a world leader; it accounts for 60 per cent of global solar cell manufacturing capacity annually, and the state uses about half of all global solar power generated per year. In 2017 alone, China installed at least 50 gigawatts of solar power capacity, which exceed the targeted 105 gigawatts that the 13th Five-Year Plan (2016–2020) projected by 2020 (Buckley, Nicholas & Brown 2018). At the same time, in terms of wind power, abundant wind energy resources in China have motivated the Chinese to develop wind power on a large scale (Sahu 2018). In 2016, China had developed 35 per cent of all cumulative wind power installations worldwide. The objective of the 13th Five-Year Plan (2016–2020) for wind power is at least 210 gigawatts, and China is expected to achieve that target by 2019 (International Energy Agency 2017).

In addition to renewable energy sources, nuclear power plays an important role in China's plans to reduce the share of fossil fuels in its energy mix. Since the 2000s, the number of nuclear power plants has increased more than tenfold in China, and in January 2018, mainland China had 38 nuclear plants in operation and about 20 more under construction (World Nuclear Association 2018). After the Fukushima Daichi nuclear disaster in Japan in 2011, the construction of new nuclear power plants slowed as the Chinese government took 'timely, cautious and comprehensive' measures to ensure the safety of nuclear plants, including those under construction (Hubbard 2014). Also in response, the State Council issued new nuclear safety guidelines consistent with the standards of the International Atomic Energy Agency (ibid.). Following safety checks, the construction of new nuclear power plants resumed, and the Chinese government continues to regard nuclear power as an essential part of the state's energy mix (Zhang & Zhao 2013).

Regarding automotive emissions, China counted 200 million motor vehicles in its population in March 2017. Despite the massive total, the figure indicated only 125 vehicles per 1,000 people in China according to the 2015 census; by comparison, some developed countries count 300–500 vehicles per 1,000 people (Hao et al. 2017). Given China's burgeoning young population, the number of cars is expected to increase rapidly in the coming years, which is slated to cause severe air pollution locally and hamper China from reaching peak carbon emissions by 2030. In response, because electric vehicles are considered to be a worthwhile alternative to traditional motor vehicles, China's government has issued several plans and policies to promote their development, including 2012's 'Energy Saving and New Energy Automotive Industry Development Plan (2012–2020)' and 2014's 'Guiding Opinions on Accelerating the Popularization and Application of New Energy Vehicles' (Chen et al. 2017). Likewise, the government took a major step in September 2017 by announcing a forthcoming ban on the production and sale of fossil fuel-powered cars (Xinhua 2017), which will no doubt urge the development of hybrid and electric vehicles not only in China but also around the world.

In December 2017, the Chinese government took another important step towards realising a low-carbon future by announcing that the entire power generation sector in China, which produces nearly half of the country's greenhouse gas emissions, will be covered by a nationwide carbon trading system in the future (Xu & Mason 2017). The system will be the largest in the world and oversee more carbon than the EU cap-and-trade system. In effect, the announcement sent a strong signal to companies worldwide that, in the future, successful business will be low carbon. Moreover, China demonstrates outstanding potential for carbon capture and storage (Dahowski et al. 2009) and is now the world's largest investor in such technology (Garnaut 2013).

Of course, China's current energy policy is also significantly influenced by the One Belt, One Road initiative. To some extent, this initiative as well as China-led financial institutions, including the Asian Infrastructure Investment Bank and the New Development Bank, invest in green technologies in Eurasia and beyond. However, the initiative is technology-agnostic; it not only pursues top advancement of green technology but also invests in old technologies (Buckley, Nicholas & Brown 2018). Since the government has restricted Chinese companies from developing new coal-fired plants at home, they have invested in such projects overseas, especially in developing countries. For example, three quarters of new energy generated by the China–Pakistan Economic Corridor, which includes 19 energy projects in renewable and coal-fired power plants, transmission and other infrastructure, comes from coal (Shaikh & Tunio 2017). Moreover, in Kenya, a Chinese company plans to construct a coal-fired power plant adjacent to a UN Educational, Scientific and Cultural Organization heritage site despite environmental and social concerns (Wesangula 2017). Although the Chinese government cannot be held accountable for all actions taken by Chinese companies, because the projects in Pakistan and Kenya are tied to the One Belt, One Road initiative they are presumably linked to the state's grand economic strategy.

Last, China supports a vast research programme on geoengineering,[5] which has been characterised as a cost-effective way to respond to climate change. Although several geoengineering technologies have been proposed, it remains unclear how well they would work and whether they would cause any (environmentally) harmful side effects. Morally speaking, geoengineering is a contested field (e.g. Gardiner 2010; Barrett 2008, 51). Stephen Gardiner (2011, 345), for example, warns that the current generation and especially citizens of affluent countries 'should be especially cautious about arguments that appear to diminish our moral responsibilities'. From that perspective, it would therefore be 'better if countries could commit themselves not to resort to geoengineering' because then the 'world would have no alternative but to reduce emissions' (Barrett 2008, 46). Geoengineering is also a controversial field from a political standpoint. As Scott Barrett (2008, 53) puts it, the problem with geoengineering is '"not how to get countries to do it' but how 'to address the fundamental question of who should decide whether and how geoengineering should be attempted – a problem of governance"'. In other words, should geoengineering projects be executed multilaterally, or should states act unilaterally, even if their projects could alter the living conditions in other countries (cf. Steffen et al. 2011, 752)? In any case, because the potential for geoengineering is enormous, the rules of geoengineering practices urgently need to be negotiated or commitments made to refrain from geoengineering.

To that end, a secondary institution is perhaps necessary to govern geoengineering practices in the future.

Adaptation to climate change

Climate change clearly poses a significant threat to human security around the world. Measured in terms of human suffering, China has faced huge losses; from 1980 to 2002, nearly 50 million Chinese lost their homes and nearly 45,000 Chinese were killed due to climate-related disasters (Roberts & Parks 2007, 76). Despite the obvious difficulty of estimating the extent to which climate change is responsible for past climate-related disasters, such figures reveal that China is located in an area exceptionally vulnerable to such disasters, the frequency and intensity of which are predicted to increase due to climate change. The Chinese government has been increasingly aware of the adverse effects of climate change in China, and in 2006, Chinese scientists published the first National Assessment Report on Climate Change, after which China's first National Climate Change Programme recognised in 2007 that the climate was already changing in China. Not only had the average surface temperature already increased 0.5–0.8°C during the twentieth century but mountain glaciers were also melting at an accelerated rate, the frequency and intensity of heatwaves had increased in the northern provinces and heavy precipitation had increased in the southern ones. In addition, many Chinese rivers had dried up due to accelerated economic and population growth, and even more critical water shortages can be expected in the future, meaning decreased agricultural output, particularly in China's northern provinces. The National Climate Change Programme therefore acknowledged that though it is essential to 'place equal emphasis on both mitigation and adaptation', given China's status as a developing country, adaptation is a 'more present and imminent task' than mitigation (National Development and Reform Commission 2007). Since then, China's government has sought to enhance the country's capacity to adapt to climate change by integrating corresponding plans into China's overall development policies.

In 2013, a year after publishing the 710-page 'Second National Assessment Report on Climate Change', the Chinese government issued its first nationwide climate change adaptation strategy, which warned all of Chinese society that is was poorly prepared to face serious threats posed by climate change. The strategy informed China's population that the government 'attaches great importance to climate change adaptation by integrating it into plans to develop the national economy and society' (National Development and Reform Commission 2013, my translation). Taking 'significantly enhanced adaptation capacity' as the ultimate goal, the plan outlined a wide range of measures to be implemented by 2020 in order to protect water, forest and soil resources, safeguard agricultural output, strengthen infrastructure, improve risk management systems, increase public awareness and establish institutional mechanisms (ibid.). Furthermore, China's Nationally Determined

Contribution to the UNFCCC announced in 2015 that the state would make efforts to adapt to climate change

> by enhancing mechanisms and capacities to effectively defend against climate change risks in key areas such as agriculture, forestry and water resources, as well as in cities, coastal and ecologically vulnerable areas and to progressively strengthen early warning and emergency response systems and disaster prevention and reduction mechanisms.
> (National Development and Reform Commission 2015)

That same year, China published the 900-page 'Third National Assessment Report on Climate Change' following a three-year compilation process involving approximately 550 scientists and experts. According to the report, average temperatures across China have increased by 0.9–1.4°C since 1909, while the coastal sea level has by risen 2.9 mm annually between 1980 and 2012. Both estimates suggest that the climate is changing at a faster rate in China than globally. The report projects that by the end of the twenty-first century, average temperatures will have risen 1.3–5.0°C in China, even if the global goal of limiting global temperature increases to 2°C is met (China Climate Change Info-Net 2015). In particular, many of China's metropoles located in coastal areas, including Shanghai, Tianjin and Hong Kong, are at high risk of flooding due to rising sea levels. It is thus argued that China will suffer the most from the business-as-usual trend of global warming and could gain the most from limiting warming to 2°C (Strauss, Kulp & Levermann 2015, 10). Such findings doubtlessly provide the Chinese government with strong domestic incentives to curb emissions and pay close attention to adapting China to climate change.

Conclusion

China's participation in international climate politics is motivated by both calculations of self-interest and beliefs about its identity as a nation. Traditionally, China's chief interests in international climate politics have included the protection of its sovereignty and the promotion of its economic development by receiving Clean Development Mechanism projects, technology transfer and other forms of foreign assistance. In China's domestic development plans, however, environmental protection has begun to receive greater consideration than ever before. Given Chinese beliefs that technological progress combined with capitalist efficiency, demand and motivation can solve all environmental problems, green technology and nuclear power play a key role in China's climate policy. Although adapting to climate change also plays an important part, it remains an underdeveloped dimension of China's domestic climate responsibility. On the international stage, image politics has also been an important driver of China's participation in international climate politics in the 2010s, and since the election of Donald Trump as US president,

China has increasingly portrayed itself as a global climate leader. However, China's growing emphasis on environmental responsibility has not translated into making ambitious commitments in international climate negotiations. Although it has committed to reach peak carbon emissions by 2030, the Chinese government continues to refuse to sign a legally binding carbon emissions reductions target, even despite strong domestic incentives to decrease the use of coal and to invest in clean technologies. By responding to such incentives and upholding the Paris Agreement, the Chinese government could not only assure its citizens that it takes their environmental concerns and future prosperity seriously but also demonstrate to the world that it is fit to take an active role in international efforts to mitigate climate change.

Notes

1 For an overview of China's environmental problems, see Kassiola and Guo (2010).
2 In Chinese culture, the concept of face is often used to describe human concerns over honour, respect or the image of oneself presented to others.
3 For further information, see the National Development and Reform Commission (2014) and annual white papers on China's climate policies and actions issued by the commission.
4 In accordance with the official translation of 生态文明, the English version of the plan should be called the 'Integrated Reform Plan for Promoting Ecological Civilization'.
5 According to Scott Barrett's (2008, 45) definition, *geoengineering* means 'the deliberate modification of the climate by means other than by changing the atmospheric concentration of greenhouse gases'.

Bibliography

Barrett, Scott. 2008. 'The incredible economics of geoengineering'. *Environmental and Resource Economics* 39:1, 45–54.

Brady, Anne-Marie. 2012. 'The Beijing Olympics as a campaign of mass distraction'. In Brady, Anne-Marie (ed.), *China's Thought Management*. New York: Routledge, 11–35.

Buckley, Tim, Simon Nicholas & Melissa Brown. 2018. 'China 2017 Review. World's Second-Biggest Economy Continues to Drive Global Trends in Energy Investment'. Institute for Energy Economics and Financial Analysis, Cleveland, OH.

Buzan, Barry. 2004. *From International to World Society? English School Theory and the Social Structure of Globalisation*. Cambridge: Cambridge University Press.

Chen, Zhihao, Linwei Ma, Pei Liu & Zheng Li. 2017. 'Electric vehicle development in China: A charging behavior and power sector supply balance analysis'. *Chemical Ingineering Research and Design* (article in press). DOI: 10.1016/j.cherd.2017.11.016.

China Climate Change Info-Net. 2015. '《第三次气候变化国家评估报告》发布' [The 'Third National Climate Change Assessment Report' released]. Accessed 4 March 2016. www.ccchina.gov.cn/Detail.aspx?newsId=56949&TId=57.

Climate Change Tracker. 2018. 'Country summary'. Accessed 4 June 2018. http://climateactiontracker.org/countries/china.html.

Dahowski, R. T. et al. 2009. 'A preliminary cost curve assessment of carbon dioxide capture and storage potential in China'. *Energy Procedia* 1:1, 2849–2856.

Deng, Yong. 2008. *China's Struggle for Status: The Realignment of International Relations.* Cambridge: Cambridge University Press.

Dent, Christopher M. 2014. *Renewable Energy in East Asia: Towards a new developmentalism.* New York: Routledge.

Ding, Sheng. 2008. 'To build a "harmonious world": China's soft power wielding in the Global South'. *Journal of Chinese Political Science* 13:2, 193–213.

Economy, Elizabeth. 1998. 'China's environmental diplomacy'. In Samuel S. Kim (ed.), *Chinese Foreign Policy Faces the New Millennium*, 4th edition. Boulder, CO: Westview Press, 264–283.

Fan, Cindy C. 2006. 'China's Eleventh Five-Year Plan (2006–2010): From "getting rich first" to "common prosperity"'. *Eurasian Geography and Economics*, 47:6, 708–723.

Fu, Sha, Zou Ji & Linwei Liu. 2015. '对中国国家自主贡献的几点评论' [Comments on China's intended nationally determined contribution]. China's National Center for Climate Change Strategy and International Cooperation, Beijing.Accessed 4 March 2016. http://files.ncsc.org.cn/www/201506/20150630222928152.pdf.

Gardiner, Stephen M. 2011. *A Perfect Moral Storm: The Ethical Tragedy of Climate Change.* Oxford: Oxford University Press.

Gardiner, Stephen M. 2010. 'Is "arming the future" with geoengineering really the lesser evil?'. In Stephen M. GardinerSimon Caney, Dale Jamieson & Henry Shue (eds), *Climate Ethics: Essential Readings.* Oxford: Oxford University Press, 284–312.

Garnaut, Ross. 2013. 'China's contribution to the global mitigation effort'. *East Asia Forum*, 26 June. Accessed 3 March 2018. www.eastasiaforum.org/2013/06/26/chinas-contribution-to-the-global-mitigation-effort.

Hao, Han, Xiang Cheng, Zongwei Liu & Fuquan Zhao. 2017. 'Electric vehicles for greenhouse gas reduction in China: A costeffectiveness analysis'. *Transportation Research* (Part D: Transport and Environment) 56, 68–84.

Harrington, Jonathan. 2005. '"Panda diplomacy": State environmentalism, international relations and Chinese foreign policy'. In Paul. G. Harris (ed.), *Confronting Environmental Change in East and Southeast Asia: Eco-Politics, Foreign Policy, and Sustainable Development.* London: United Nations University Press andEarthscan, 102–118.

Harris, Paul G. 2013. *What's Wrong with Climate Politics and How to Fix It.* Cambridge: Polity Press.

Heggelund, Gørild, Steinar Andresen & Inga Fritzen Buan. 2010. 'Chinese climate policy: Domestic priorities, foreign policy, and emerging implementation'. In Kathryn Harrison & Lisa McIntosh Sundstrom (eds), *Global Commons, Domestic Decisions.* Cambridge, MA:MIT Press. 229–259.

Hopwood, Bill, Mary Mellor & Geoff O'Brien. 2005. 'Sustainable development: Mapping different approaches'. *Sustainable Development* 13, 38–52.

Hu, Jintao. 2007. 'Hold high the great banner of socialism with Chinese characteristics and strive for new victories in building a moderately prosperous society in all respects'. Accessed 4 March 2018. www.china.org.cn/english/congress/229611.htm.

Hubbard, Christopher. 2014. *Fukushima and Beyond: Nuclear Power in a Low-Carbon World.* New York: Routledge.

Hurd, Ian. 2007. *After Anarchy: Legitimacy and Power in the United Nations Security Council.* Princeton: Princeton University Press.

Hurd, Ian. 1999. 'Legitimacy and authority in international politics'. *International Organization* 53:2, 379–408.

Hurrell, Andrew. 2007. *On Global order. Power, Values, and the Constitution of International Society.* Oxford: Oxford University Press.

Höhne, Niklas, Thomas Day, Gesine Hänsel & Hanna Fekete. 2015. 'Assessing the missed benefits of countries' national contributions'. NewClimate Institute for Climate Policy and Global Sustainability, Cologne.

Information Office of the State Council of the People's Republic of China. 2011. 'White paper: China's policies and actions for addressing climate change'. Accessed 3 March 2018. www.china-un.org/eng/hyyfy/t521513.htm.

Information Office of the State Council of the People's Republic of China. 2008. 'White paper: China's policies and actions for addressing climate change'. Accessed 3 March 2018. www.china-un.org/eng/hyyfy/t521513.htm.

International Energy Agency. 2017. 'Renewables 2017: Analysis and forecasts to 2022. Executive summary'. Accessed 3 February 2018. www.iea.org/Textbase/npsum/renew2017MRSsum.pdf.

International Energy Agency. 2014. 'Energy subsidies'. Accessed 4 March 2016. www.worldenergyoutlook.org/resources/energysubsidies/

Johnston, Alistair I. 1998: 'China and international institutions: A decision rule analysis'. In Michael B. McElroy, Chris P. Nielson & Peter Lydon (eds), *Energizing China: Reconciling Environmental Protection and Economic Growth.* Cambridge, MA: Harvard University Press, 555–599.

Kassiola, Joel Jay & Sujian Guo (eds). 2010. *China's Environmental Crisis: Domestic and Global Political Impacts and Responses.* New York: Palgrave Macmillan.

Kobayashi, Yuka. 2005. 'The 'troubled modernizer': Three decades of Chinese environmental policy and diplomacy'. In Paul G. Harris (ed.), *Confronting Environmental Change in East and Southeast Asia: Eco-Politics, Foreign Policy, and Sustainable Development.* London: United Nations University Press and Earthscan, 87–101.

Kong, Yigang, Jie Wang, Zhigang Kong, Furong Song, Zhiqi Liu & Congmei Wei. 2015. 'Small hydropower in China: The survey and sustainable future'. *Renewable and Sustainable Energy Reviews* 48, 425–433.

Kong, Yigang, Zhigang Kong, Zhiqi Liu, Congmei Wei & Gaocheng An. 2016. 'Substituting small hydropower for fuel: The practice of China and the sustainable development'. *Renewable and Sustainable Energy Reviews* 65, 978–991.

Kopra, Sanna. 2016. 'Great power management and China's responsibility in international climate politics'. *Journal of China and International Relations* 4:1, 20–44.

Kopra, Sanna. 2012. 'Is China a responsible developing country? Climate change diplomacy and national image building'. Centre for Qualitative Social Research, Department of Sociology, Hong Kong Shue Yan University, Social and Cultural Research Occasional Paper no 13.

Lan, Lan. 2010. '"No intention" of capping emissions'. *China Daily*, 25 February. Accessed 3 March 2018. www.chinadaily.com.cn/china/2010-02/25/content_9499066.htm.

Li, Keqiang. 2014. 'Report on the work of the government'. Accessed 20 February 2015. http://news.xinhuanet.com/english/special/2014-03/14/c_133187027.htm.

Li, Peng. 1995. 'Report on the outline of the Ninth Five-Year Plan (1996–2000) for National Economic and Social Development and the Long-range Objectives to the Year 2010 (excerpts)'. Accessed 20 February 2015. www.china.org.cn/95e/95-english1/2.htm.

Li, Xiao-zhu, Zhi-jun Chen, Xiao-chao Fan & Zhi-jiang Cheng. 2018. 'Hydropower development situation and prospects in China'. *Renewable and Sustainable Energy Reviews* 82, 232–239.

Liu, Coco & ClimateWire. 2014. 'China will limit pollution from steel and cement'. *Scientific American*, 6 November. Accessed 22 February 2017. www.scientificamerican.com/article/china-will-limit-pollution-from-steel-and-cement/

Ministry of Environmental Protection. 2008. 'National Eleventh Five-Year Plan for Environmental Protection'. Accessed 22 February 2017. http://english.mep.gov.cn/Plans_Reports/11th_five_year_plan/200803/t20080305_119001_1.htm.

National Development and Reform Commission. 2015. 'Enhanced actions on climate change: China's intended nationally determined contribution'. Accessed 4 March 2018. http://www4.unfccc.int/submissions/indc/Submission%20Pages/submissions.aspx

National Development and Reform Commission. 2014. '国家应对气候变化规划 (2014–2020年)' [National Climate Change Plan (2014–2020)]. Accessed 22 February 2017. www.ccchina.gov.cn/nDetail.aspx?newsId=49211&TId=60.

National Development and Reform Commission. 2013. '国家适应气候变化战略' [National Climate Change Adaptation Strategy]. Accessed 22 February 2017. www.sdpc.gov.cn/zcfb/zcfbtz/201312/W020131209343322750059.pdf

National Development and Reform Commission. 2012. 'China's policies and actions for addressing climate change'. Accessed 22 February 2017. www.ccchina.gov.cn/WebSite/CCChina/UpFile/File1324.pdf.

National Development and Reform Commission. 2007. 'China's National Climate Change Programme'. Accessed 22 February 2017. www.china.org.cn/english/environment/213624.htm.

National People's Congress & Chinese People's Political Consultative Conference. 2016. 中华人民共和国国民经济和社会发展第十三个五年规划纲要 [The Thirteenth Five-Year Plan for the National Economic and Social Development of the People 's Republic of China]. Accessed 23 October 2016. http://news.xinhuanet.com/politics/2016lh/2016-03/17/c_1118366322.htm.

Pan, Jihua, Haibing Ma & Ying Zhang. 2011. 'Green economy and green jobs in China: Current status and potentials for 2020'. Worldwatch Report 185.

Percival, Bronson. 2007. *The Dragon Looks South: China and Southeast Asia in the New Century*. Westport: Praeger Security International.

Reuters. 2016. 'China to cap coal at 55 percent of total power output by 2020: NEA'. 7 November. Accessed 22 February 2018. www.reuters.com/article/us-china-power-consumption/china-to-cap-coal-at-55-percent-of-total-power-output-by-2020-nea-idUSKBN1320LT.

Roberts, J. Timmons & Bradley C. Parks. 2007. *A Climate of Injustice: Global inequality, North-South Politics, and Climate Policy*. Cambridge, MA, and London: MIT Press.

Sahu, Bikash Kumar. 2018. 'Wind energy developments and policies in China: A short review'. *Renewable and Sustainable Energy Reviews* 81:1, 1393–1405.

Shaikh, Saleem & Sughra Tunio. 2017. 'Pakistan ramps up coal power with Chinese-backed plants'. Reuters, 3 May. Accessed 22 February 2018. www.reuters.com/article/us-pakistan-energy-coal/pakistan-ramps-up-coal-power-with-chinese-backed-plants-idUSKBN17Z019.

Shouqiu, Cai and Mark Voigts. 1993. 'The development of China's environmental diplomacy'. *Pacific Rim Law & Policy Journal* 3, 17–42.

Standing Committee of the National People's Congress. 2009. '全国人民代表大会常务委员会关于积极应对气候变化的决议'. [Standing Committee of the National People's Congress resolution on tackling climate change]. Accessed 23 October 2016. www.npc.gov.cn/npc/xinwen/rdyw/wj/2009-08/27/content_1516165.htm.

Steffen, Will, Åsa Persson, Lisa Deutsch, Jan Zalasiewicz, Mark Williams, Katherine Richardson, Carole Crumley, Paul Crutzen, Carl Folke, Line Gordon, Mario Molina, Veerabhadran Ramanathan, Johan Rockström, Marten Scheffer, Hans Joachim Schellnhuber & Uno Svedin. 2011. 'The Anthropocene: From global change to planetary stewardship'. *AMBIO* 40(7), 739–761.

Stern, Nicholas. 2006. 'Stern Review: The economics of climate change'. Accessed 23 October 2016. http://webarchive.nationalarchives.gov.uk/+/http://www.hm-treasury.gov.uk/stern_review_report.htm.

Strauss, Benjamin H., Scott Kulp & Anders Levermann. 2015. 'Mapping choices: Carbon, climate, and rising seas, our global legacy'. *Climate Central Research Report*, 1–38. Accessed 2 May 2018. http://sealevel.climatecentral.org/uploads/research/Global-Mapping-Choices-Report.pdf.

Su, Wei. 2010. 'Untitled letter from Su Wei, Director-General of Department of Climate Change, National Development and Reform Commission of China to Yvo de Boer, Executive Secretary of the UNFCCC Secretariat'. Accessed 23 October 2016. https://unfccc.int/files/meetings/cop_15/copenhagen_accord/application/pdf/chinacphaccord_app2.pdf.

UNFCCC (United Nations Framework Convention on Climate Change). 2015. 'Adoption of the Paris Agreement: Proposal by the president', 12 December. Accessed 14 March 2017. http://unfccc.int/documentation/documents/advanced_search/items/6911.php?priref=600008831..

UNFCCC (United Nations Framework Convention on Climate Change). 1992. 'United Nations Framework Convention on Climate Change'. Accessed 3 March 2017. http://unfccc.int/files/essential_background/background_publications_htmlpdf/application/pdf/conveng.pdf.

U.S.–China Climate Leaders' Declaration. 2015. 'On the occasion of the first session of the U.S.-China Climate-Smart/Low-Carbon Cities summit Los Angeles, CA, September 15–16, 2015.' Accessed 3 March 2017. http://go.wh.gov/cSqu8N.

Wang, Hongying. 2005. 'National image building and Chinese foreign policy'. In Yong Deng & Fei-Ling Wang (eds), *China Rising. Power and Motivation in Chinese Foreign Policy*, Lanham: Rowman & Littlefield Publishers, 73–102.

Wendt, Alexander. 1999. *Social Theory of International Politics.* Cambridge: Cambridge University Press.

Wesangula, Daniel. 2017. 'Kenya signs China deal for coal plant beside Unesco site'. Climate Home News, 23 May. Accessed 22 February 2018. www.climatechangenews.com/2017/05/23/kenya-signs-china-deal-coal-plant-beside-unesco-site/.

White House. 2015a. 'U.S.-China joint presidential statement on climate change'. Accessed 3 March 2017. www.whitehouse.gov/the-press-office/2015/09/25/us-china-joint-presidential-statement-climate-change.

White House. 2015b. 'Fact sheet: U.S.–China climate leaders summit'. Accessed 3 March 2017. www.whitehouse.gov/the-press-office/2015/09/15/fact-sheet-us-%E2%80%93-china-climate-leaders-summit.

White House. 2014. 'U.S.-China joint announcement on climate change'. Accessed 3 March 2017. www.whitehouse.gov/the-press-office/2014/11/11/us-china-joint-announcement-climate-change

World Nuclear Association. 2018. 'Nuclear power in China'. Accessed 22 February 2018. www.world-nuclear.org/information-library/country-profiles/countries-a-f/china-nuclear-power.aspx.

Wu, Xinbo. 2001. 'Four contradictions constraining China's foreign policy behavior'. *Journal of Contemporary China* 10:27, 293–301.

Xie, Zhenhua. 2010. 'Speech at the high level segment of COP16&CMP6'. Accessed 23 October 2016. http://unfccc.int/files/meetings/cop_16/statements/application/pdf/101208_cop16_hls_china.pdf.

Xinhua. 2017. 'Economic Watch: China mulls timetable to ban fossil fuel vehicles'. 11 September. Accessed 22 February 2018. www.xinhuanet.com/english/2017-09/11/c_136601024.htm.

Xinhua. 2015a. '授权发布 中共中央 国务院关于加快推进生态文明建设的意见' [Authorized release: An opinion of the Chinese Communist Party Central Committee and State Council regarding acceleration of promotion of the construction of ecological civilization]. 5 May. Accessed 23 October 2016. http://news.xinhuanet.com/politics/2015-05/05/c_1115187518.htm

Xinhua. 2015b. 'Full Text: Integrated reform plan for promoting ecological progress'. 21 September. Accessed 23 October 2016. http://news.xinhuanet.com/english/china/2015-09/21/c_134646023.htm.

Xinhua. 2014. 'China unveils energy strategy, targets for 2020'. 19 November. Accessed 23 October 2016. http://news.xinhuanet.com/english/china/2014-11/19/c_133801014.htm.

Xinhua. 2009. 'China announces targets on carbon emission cuts'. 26 November. Accessed 23 October 2016. http://news.xinhuanet.com/english/2009-11/26/content_12544181.htm.

Xu, Muyu, & Josephine Mason. 2017 'China aims for emission trading scheme in big step vs. global warming'. Reuters, 19 December. Accessed 22 February 2018. www.reuters.com/article/us-china-carbon/china-aims-for-emission-trading-scheme-in-big-step-vs-global-warming-idUSKBN1ED0R6.

Zhang, Hui & Shangui Zhao. 2013. 'China moves cautiously ahead on nuclear energy'. Bulletin of the Atomic Scientists, 22 April. Accessed 22 February 2018. https://thebulletin.org/china-moves-cautiously-ahead-nuclear-energy

Zhao, Xingang et al. 2012. 'A critical-analysis on the development of China hydropower'. *Renewable Energy* 44, 1–6.

7 Great climate irresponsibles?[1]

During the past 40 years, as environmental norms have slowly but steadily been integrated into international society, profound normative change has occurred as states have created a complex set of rules and institutions to sustain and organise environmental practices. In that sense, the answer to Robert Jackson's (1996) question 'Can international society be green?' seems to be positive; many environmental treaties and organisations enjoy almost universal support, and states, including great powers, routinely participate in environmental diplomacy. International society has also succeeded in developing important rules, including common but differentiated responsibilities, for the distribution of special climate responsibilities among states. In another sense, as environmental degradation persistently accumulates and climate change accelerates, it remains frustratingly difficult to give a favourable answer to Jackson's question. After all, international society has failed to promote genuine international justice, real human suffering has not been alleviated, and social disparities continue to compound both locally and globally. In response to those failures, the need to transform international climate practices remains critical. In this chapter, I summarise the chief contributions of the book, as well as discuss weaknesses in the international norms of climate responsibility. I argue that a more balanced, solidarist conceptualisation of the human–nature relationship is necessary to achieve the genuine change required to make international society 'green'. From that perspective, given their role as gatekeepers of international society, great powers are bound to shoulder more responsibility for mitigating and adapting to climate change.

Climate responsibility in international society

In this book, I have examined the institutionalisation of international norms of climate responsibility via the lens of English School theory. Traditionally, English School scholars have been largely uninterested in secondary institutions (i.e. international organisations and regimes) and tended to believe that regime theorists are responsible for studying instrumental secondary institutions. Recently, however, a growing number of English School theorists have stated that the holistic approach of the English School, which primarily focuses on primary institutions in institutional change, is necessary to comprehend the evolution of international norms and practices but cannot afford complete understanding on its own. In contrast to liberal institutionalists, who assume that international organisations create and manage international order by formulating international rules and procedures, English School theorists emphasise that international order is organised and sustained by primary institutions, which also shape the operations of secondary ones. When it comes to international responsibilities, however, the role of secondary institutions is vital. Without the establishment of secondary institutions, it would be difficult, if not impossible, to negotiate and mediate the positive responsibilities of states concerning global problems. As I have illustrated, international responsibilities are not given; on the contrary, participants in secondary institutions negotiate the responsibilities of themselves and other participants, the ethical grounds on which such responsibilities are distributed among participants and by which mechanisms they are implemented and monitored. However, primary institutions always shape those negotiations as well. Although some responsibilities are formulated as legal obligations in international treaties, most state responsibilities are uncodified in international law and thus remain informal.

Given my interest in the evolution of international norms and practices, I have traced the early development of rules pertaining to the norm of climate responsibility at length. In so doing, I have sought to denaturalise assumptions related to climate diplomacy. International norms and practices do not emerge out of thin air, nor can they be traced to a single treaty. Although the 1992 UN Conference on Environment and Development in Rio de Janeiro marked a watershed for the institutionalisation of environmental norms, it would be a mistake to assume that practices of climate responsibility began to evolve there. As I have shown, to fully trace the evolution of climate responsibility, we need to delve further into the past and analyse the myriad international environmental treaties and conferences since the 1970s, if not before. As the concept of climate responsibility was constructed, the cornerstones were laid in 1972, well before the concept's broader architecture was added with the formulation of the UN Framework Convention on Climate Change (UNFCCC) in 1992. To trace the concept's evolution in China in particular, I also briefly explored ancient China's and Maoist China's environmental practices, which form the ideological basis for contemporary ideas of climate responsibility in China. As I have demonstrated, China's stance during international environmental negotiations since 1972 has not significantly changed. Indeed, as my investigation into the history of climate responsibility has revealed, isolated Maoist China was not a norm-taker but a norm-maker in international environmental practices. As such, China, amid its rise in the twenty-first century, will likely be a status quo power in international environmental politics.

At present, climate responsibility is a highly state-centric norm. Above all, it focuses on interstate responsibilities – that is, states' obligations to reduce emissions and provide assistance to poorer states – and does not pay much attention to cosmopolitan notions of responsibility (cf. Harris 2010). Although the norm takes an unusually anthropocentric approach to nature, it is not a characteristically humanist norm, for it does not clearly recognise the environmental rights of people. At the same time, and perhaps needless to say, climate responsibility does not position nature as its moral referent object and largely dismisses its rights as a result. Nevertheless, the emergence of the international norm of climate responsibility has spelled profound normative change in international society, in which environmental stewardship has come to constitute a collective responsibility of both states and non-state actors (cf. Falkner 2012; Falkner and Buzan 2018). Climate responsibility has thus clearly become an established international norm with which even the most powerful states must comply if they wish to be and be seen as responsible international agents (Kopra forthcoming). Consequently, states have continued to participate in climate negotiations since the development of the Kyoto Protocol even if they initially disapproved or later withdrew from the agreement. Despite widespread, continued conflict regarding climate responsibility, as well as pointed criticism of the contributions of other states, no state has simply walked away from climate negotiations.

The UNFCCC functions as the single most important secondary institution regarding climate responsibility. Nevertheless, in being so quickly negotiated – in about two years – the treaty has promoted two assumptions regarding the governance of climate responsibility. On the one hand, its rapid formation suggested both that states regarded climate change as a common concern of international society and that a correspondingly strong political will existed to take action against it. On the other, the UNFCCC was not initially characterised as a strong organisation that could harm a state's sovereignty or national interests (Kopra 2018, Kopra forthcoming). Consequently, the UNFCCC has failed to cultivate a so-called 'thick' international society around climate responsibility, and many central questions related to such responsibility, including finance and compensation, remain unanswered (cf. Palmujoki 2013, 191–192). Moreover, apart from general ideas of state responsibility, the UNFCCC imposes no rules for liability in causing environmental harm, even if compensation for damage caused to future generations or Earth itself could be calculated and given force in the first place. Despite the UNFCCC's failure to decelerate climate change, it clearly remains significant in international society. In general, secondary institutions are important because they provide a social forum in which states and non-state actors can sustain a dialogue, which, at best, can increase their and the general public's environmental awareness and willingness to strengthen the rules of environmental governance. In other words, the UNFCCC creates a political framework in which state and non-state actors can negotiate more effective and ambitious climate-related agreements and thus improve their future performance (Kopra 2018, Kopra forthcoming). If the UNFCCC were to be dismantled, no alternative, equally effective forum for advocating international commitments would exist, and there is no time to establish another in its stead. Nevertheless, the UNFCCC has no intrinsic value and could – some may say it should – be displaced by post-national climate practices in the future.

Of course, the UNFCCC is no longer the sole platform for international climate action. A growing number of non-state climate change initiatives and experiments take root year after year, which could signal emerging notions of cosmopolitan climate responsibility. On the one hand, the number of nongovernmental climate initiatives and organisations indicate the UNFCCC's weakness in resolving the climate crisis and the failure of states to shoulder their climate responsibility, both of which signs call for alternative approaches. On the other, the active participation of non-state actors demonstrates that world society is convinced of the urgency of climate change and that, without their participation, efforts to mitigate climate change are likely to fail. From the perspective of the English School, the potential fragmentation and diversification of climate governance is not necessarily a weakness but 'indicates a common understanding and a "thick" interpretation' of climate responsibility (Palmujoki 2013, 195). From that viewpoint, private and public initiatives outside the UNFCCC not only offer multiple, potentially more efficient channels to organise and enact climate responsibility but also engage the broader participation of both state and non-state actors (ibid., 192). That view also poses the critical question of whether climate responsibility will develop into a so-called 'standard of civilization' that defines and validates the practices of civilized members of international society and world society in the future (Kopra forthcoming).

Some observers argue that sovereignty stands as the greatest obstacle to effective environmental protection and that transcending state-centric international society is thus a precondition for the genuine assimilation of climate responsibility (e.g. Falk 1972; Harris 2010). However, sovereignty *per se* is not the principal reason why international society has failed to respond to climate change and other global environmental problems, for any other universal political order, whether some sort of world government or cosmopolitan world society, would face the same problems. Moreover, no evidence suggests that such a world government or society would somehow be more solidarist or ecocentric than today's state-centric one, and in any case, it would be similarly shaped by the power struggles and calculations of self-interest among individuals (cf. Williams 2005). Without a minimum international order in place, responding to climate change and organising action against it would be exceedingly difficult. If relatively small units such as states find it challenging to agree on effective political actions, then how would larger units be able to make any difference?

Arguably, sovereignty has to be redefined in more solidarist terms in order to function in today's global era; it should not only focus on inter-state relations and justice but also pay more attention to the wellbeing of people. Accordingly, I conclude that a key reason for the failure of international society to find effective and fair solutions to climate change is a lack of great power leadership. Great powers have the greatest responsibility for sustaining and organising international society, and climate change has emerged as an overriding risk to international security as well as human justice and wellbeing. Although both the United States and China have acknowledged the existence of great power climate responsibility, they have refused to commit to binding emissions reduction targets under the UNFCCC, and the United States in particular has failed to fulfil its domestic climate responsibility and provide developing countries with much-needed assistance to mitigate and adapt to climate change.

That said, great power management has played an important role in institutionalising the international norm of climate responsibility. Under the leadership of the United States, prominent steps were taken from the 1960s to early 1990s, during which time China shaped the construction of the building blocks of environmental stewardship. Although the United States abandoned the Kyoto Protocol and little progress was made in international climate negotiations in the early 2000s, during the presidency of Barack Obama, the United States acknowledged great power responsibility for climate change and took significant diplomatic action to promote collective efforts in response to climate change. In addition to domestic developments that elevated the position of environmental issues on the Chinese government's agenda, such action convinced China to assume a more constructive role in international climate politics. As China realised that climate responsibility could cement its membership in the great power club, it began to promote climate responsibility as an attribute of great power responsibility. In time, US–Chinese collaboration on climate change made the establishment of the Paris Agreement possible. However, insofar as both great powers have advanced voluntary domestic action plans as key means to fulfil climate responsibility, great power climate responsibility continues to be an informal international norm that will not likely be formalised in the near future, particularly as long as Donald Trump's administration continues to shirk that responsibility. Nevertheless, as long as China insists upon the link between great power status and responsibility for climate change, its emerging leadership in international climate politics could induce profound normative change in international society.

Climate responsibility with Chinese characteristics

After examining the Chinese government's notions of national responsibilities and interpretations of the international norm of climate responsibility, I argue that we cannot regard China as being irresponsible in international climate politics. After all, it has participated in international collaborations addressing climate change since the 1980s, taken more voluntary action than international law requires, and constructed a comprehensive environmental policy framework from scratch since the late 1970s. Nevertheless, China's notion of climate responsibility has remained highly retrospective by exclusively pursuing historical justice. That conceptualisation stresses the cumulative responsibility of developed countries to address climate change while allocating little responsibility to developing countries, among which China counts itself. However, China's position in international climate negotiations has changed dramatically since the 2009 UN Climate Change Conference in Copenhagen. Symptomatic of that transformation, *responsibility* has entered the party-state's lexicon, and today, Chinese political leaders and the general public widely agree that China's economic miracle has come at a high environmental price. Above all, they recognise that they – and the planet – cannot afford to follow the Western model of industrialisation based on the 'pollute first, clean up later' mentality (Xie 2010).

Because coping with environmental degradation ranks among the greatest challenges that China faces today, the Chinese government has had no choice but to integrate environmental protection, emissions reduction and energy conservation into the country's overall development targets. China's climate policy is thus largely driven by the Communist Party's concerns with its legitimacy. In terms of mitigating climate change, the motive of emissions reductions is therefore not nearly as important to China developing what it calls an 'ecological civilisation'. Only time will tell whether eco-civilisation remains primary a device of party rhetoric or exerts real political influence in transforming China's path towards development. In the meantime, however, when it comes to great power climate responsibility, drivers of climate policy *do* matter. A state can be defined as a great power only if it and other members of international society regard it as having special rights and responsibilities on the international stage (Bull 1980). By contrast, China's climate policy is chiefly motivated by domestic interests, and China can thus hardly be regarded as fulfilling great power responsibility.

Although the chief referent objects of China's climate responsibility are not humankind or the environment but the party-state and the Chinese nation, China no longer focuses solely on national responsibilities. Consequently, it seems that China increasingly identifies itself as a great power. Such a change in identity has become especially visible in the context of international financial governance, in which China has established new, alternative multilateral organisations that may challenge their US-led counterparts in the future. Because China has published all recent important climate change mitigation objectives as well as ratified the Paris Agreement in joint press conferences with the United States, it appears that it has formulated its climate policies in reference to its potential great power status. However, China continues to oppose formal discussions on climate change at the UN Security Council, largely because it does not operate under the principle of common but differentiated responsibilities, which absolves it from assuming legally binding climate responsibilities. Furthermore, China's conviction of the importance of assuring that the voices of all developing countries are heard in international climate negotiations has supported its stance that, as a decision-making body not based on universal participation, the UN Security Council is not an appropriate forum for discussing climate responsibility (cf. Wang 2011). From a realist perspective, China's motives in underscoring the UN system are clear; the Chinese government wants to preserve its privilege of having a permanent seat on the UN Security Council in order to restrain US hegemony and also support the principle of common but differentiated responsibilities, which guarantees that developing countries, including itself, need not commit to legally binding greenhouse gas emissions reductions. The Kyoto Protocol's flexible mechanisms, particularly the Clean Development Mechanism, represent important incentives for countries reluctant to take action against climate change, and China has undoubtedly benefitted a great deal from such projects. From the perspective of the English School, however, China's increasing participation in multilateral cooperation suggests that China has fully integrated into international society and may become willing to shoulder more responsibility in addressing global concerns such as climate change.

To evaluate whether China's notion of climate responsibility is ethically acceptable, it is useful to examine whether China has taken action that could be viewed as irresponsible. In other words, we can gain a better understanding of Chinese practices, perceptions and objectives by highlighting what China 'is *not* doing' (Medeiros 2009, 213). First, China does not promote its political objectives by aggressive means but participates in international multilateral negotiations. Second, it does not promote a radically alternative regional or global order but instead supports international organisations and regimes. Third, it does not pursue confrontation with the United States but seeks to actively cultivate a peaceful relationship with the United States and other great powers. Fourth, it does not dispute the reality of climate change but has integrated climate change plans into its overall policy strategies. Nevertheless, since Robert B. Zoellick's speech in 2005, Western countries began urging China to become a 'responsible stakeholder' and shoulder more global responsibility, including climate change responsibility, which indicates that China had not been a responsible stakeholder theretofore. Because, as Chen Zhimin (2009, 26) points out, 'undertaking and demanding international responsibility are both noble causes', it is not inappropriate to ask China to shoulder more global responsibility, for doing so will 'enhance the awareness of responsibility' among China's political elite and its general public. At the same time, a few important aspects should be remembered when criticising China of irresponsibility in international climate politics. First, per capita emissions are much lower in China than in the United States and many other developed countries. Second, the lion's share of China's increasing greenhouse gas emissions are caused by exports to the West, meaning that Western consumers are partly responsible for increasing emissions in China. Last, all states have national responsibilities and cannot assume international responsibilities that are beyond their economic or political abilities to bear. Indeed, all three points are central to China's stance on climate responsibility: that international expectations of its global responsibility should be closely linked to its stage of development.

China's permanent seat on the UN Security Council imposes special responsibilities upon the Chinese government. From that viewpoint, China arguably has a moral duty to bear more responsibility than smaller states and cannot escape its obligation to develop strong policies to mitigate climate change. Although China recognises that great powers have great responsibilities, it does not unequivocally define itself as a great power in the context of climate politics but describes itself as simply a responsible major or big power, mostly in reference to its size and population. Although many rural citizens remain poor in China, many urban ones are extremely wealthy and pursue lifestyles as unsustainable as those of Westerners, if not more so. At its current level of development, China's wealth and capacity to execute ambitious actions against climate change will continue to expand and make it increasingly difficult for its government to assure the world that it is a developing country. After all, if the Chinese state can afford to twice hold an extravagant Olympic Games and venture into space, then how can it not afford to mitigate climate change? Consequently, many observers have suggested that China should abandon its identity as a victim of the Century of Shame. Jin Canrong (2011, 6–7, 11–12) proposes, for instance, that China should 'establish and adjust to the great power mentality', which partly involves proactively assuming international responsibilities. Jin's conceptualisation of great power status clearly binds that status to great responsibilities: 'What's more, a country with great power mentality should be responsible and sympathetic over the weak, advocate equality and justice and work for the well-being of the people and the peace of the world' (ibid., 7). Responding to the issue of climate change could provide China with an opportunity to prove to the world its emerging status as a world leader.

Prospects for China's leadership in international climate politics

In line with China's rising international status, many observers have begun to worry whether China will cooperate in line with contemporary international norms and practices created by the United States after World War II. Although others have suggested that China will attempt to shape contemporary international society and pursue some sort of China-centred international order instead, China's increasing interest in multilateral cooperation during recent decades could signal its acceptance of contemporary international norms and its felt responsibility as a member of international society. Even if China did overtake the United States as the hegemonic leader of the world, it would face a starkly different sort of international society than previous rising powers – for example, the United States in the early twentieth century.[2] Unless world war breaks out, it is unlikely that existing secondary institutions will collapse, meaning that China will lack the opportunity to design a radically new system of global governance in accordance with its preferences. Moreover, contemporary problems differ fundamentally from those faced before and after World War II; instead of the balance-of-power security dilemmas characteristic of the Cold War period, today's problems vary from terrorism to financial crises and from environmental degradation and climate change to issues of global health and food security. No matter how powerful a state, it cannot solve those problems unilaterally but has to depend upon international cooperation. Therefore, it is unlikely, if not impossible, that China will abandon international society. In reality, voluntary participation in international organisations and compliance with international law can be important factors of a state's soft power and reputation as a good international citizen, as well as its legitimacy and credibility. Indeed, all of those values are significant drivers of Chinese politics.

Regarding climate change, David G. Victor (2011) divides the states of the world into two categories: enthusiastic and reluctant. Enthusiastic states – mostly EU countries but also essentially all developed nations – are keener to devote their resources to cutting emissions and leading international efforts against climate change. Given the President Obama's active role in international climate negotiations, Victor ranks the United States among the enthusiastic states despite its refusal to ratify the Kyoto Protocol and failure to adopt a progressive national climate policy. Today, however, led by climate-sceptic Donald Trump, the United States undoubtedly falls into the latter category – reluctant states. In 2011, Victor's reluctant country group included China and all other BRIC countries, which, despite becoming major greenhouse gas emitters, had not prioritised action against climate change on their national agendas but focused on maintaining economic growth and demanded that developed countries shoulder all responsibility for mitigating climate change. In Victor's framework, reluctant countries remain unwilling to take action against climate change unless emissions controls coincide with their national interests, such as ensuring energy security and reducing local pollution. Indeed, China long resisted taking any responsibility for mitigating climate change by instead insisting upon the historically informed responsibility of developed countries to lead those efforts, as well as refused to commit in any binding emissions reduction targets because emissions controls would have raised the cost of doing business in China. China's attitude towards allaying climate change, however, has transformed since Victor's book was published in 2011. Today, identity and belief are important drivers in shaping China's role in international climate negotiations and it seems liable to emerge as an enthusiastic country with exceptional potential to fuel a new era of international climate politics.

Most likely, China will not abandon international climate negotiations. Mitigating climate change does not contradict the objectives of the Chinese party-state, and there are strong domestic incentives to decrease the use of fossil fuels and develop green technologies instead. Moreover, since the beginning of the Trump administration, climate politics has given China an opportunity to represent itself as an international leader and make normative claims about what it means to be a responsible great power in the twenty-first century. For China, climate responsibility is thus an appealing alternative to liberal political solidarist norms of human rights as a 'new standard of civilization' and a moral foundation for great power responsibility (Kopra 2018; Kopra forthcoming). At the same time, China's emerging role as leader does not mean that it has somehow become a more solidarist international agent than before. Because the post-American era is not necessarily post-hegemonic, a China-led international society is apt to propagate a new type of hegemony (Callahan 2008). Consequently, the focal questions to address are what sort of great power responsibility China intends to shoulder, how it will demonstrate that responsibility in international affairs and what normative change its rise will generate globally.

For sceptics of technology's role in solving environmental problems, China's emerging climate leadership is not necessarily good news. Nuclear power stands as the key means to reduce China's dependency on fossil fuels, and China has invested heavily in carbon dioxide capture and storage technology. Furthermore, China's climate policy – and international climate politics in general – does not take a clear stance on the role of geoengineering in mitigating climate change, while the construction of China's massive hydropower plants have engendered numerous social problems. In short, China's domestic climate policies and human rights record do not suggest that it would be a solidarist climate leader that would promote human values globally or take a more solidarist approach at the international level. With its hard-line coalitions, China sought to re-establish the division between developed and developing countries at the 2017 UN Climate Change Conference in Bonn. Although such bifurcation is justified from the perspective of international justice and the emphasis on developed countries' pre-2020 climate commitments in Bonn was important to advance post-2020 actions against climate change, China and other major emitters arguably support bifurcation simply to avoid more ambitious emissions reduction targets for middle-income countries such as themselves.

Arguably, international society cannot be green without a major paradigm shift that promotes a new ecological consciousness and solidarist morals. State responsibility does not stand upon the number of international treaties signed but the scope and ambition of international practices, which include forming treaties. Vaclav Havel has eloquently described the need for fundamental change in contemporary international society:

> It is my deep conviction that the only option is a change in the sphere of the spirit, in the sphere of human conscience. It's not enough to invent new machines, new regulations, new institutions. We must develop a new understanding of the true purpose of our existence on this Earth. Only by making such a fundamental shift will we be able to create new models of behaviour and a new set of values for the planet.
>
> (Quoted in Speth 2008, 200)

Yet, instead of merely pointing out the need for change, we need to examine what the critical drivers of such change are. Who or what can generate the change needed for international and world society to assume climate responsibility, and how? Although I have emphasised the central role of secondary institutions in the processes of defining and distributing global responsibilities, I do not maintain that they play the key role in generating the necessary change in international society. Secondary institutions and international law represent the minimum, not the maximum, standard of conduct in international relations; moreover, they cannot guarantee that states and individuals change their environmentally harmful practices and pollutive lifestyles. Laws cannot force people to obey them, and not even authoritarian governments such as China's can achieve widespread compliance simply by issuing a policy or law. To some extent, genuine entrenchment of climate responsibility is contingent on the communiqués of global leaders and role models. From that perspective, Pope Francis's encyclical letter *Laudato si: On Care for Our Common Home*, the joint climate statement of religious and faith leaders around the world prior to the 2015 UN Climate Change Conference in Paris and the increasing number of celebrities who participate in environmental campaigns are highly auspicious signs, for they could exert significant impact on civil society's environmental awareness and discourse on climate change. In China, by analogy, former basketball player Yao Ming and film star Jackie Chan have played an influential role in WildAid's successful campaign against shark fin soup. If civil society views mitigating climate change as a moral responsibility, then it can likely influence international climate practices as well.

Based on the theoretical and empirical findings of this book, I conclude that *great power management* is an essential driver of fundamental change in international society. Accordingly, the leadership of the United States and China will be especially crucial to mobilising the political will needed to strengthen climate responsibility globally. Given President Trump's reluctance to shoulder climate responsibility, the future of great power climate responsibility looks quite grim. At the same time, China has great potential to act as a role model for climate responsibility as long as it manages to develop its sought-after ecological civilisation and modernise without recklessly increasing greenhouse gas emissions. China has already introduced new concepts of state responsibility – namely, ecological civilisation and a new type of great power relationship – both of which have the potential to transform international norms and practices. It remains unclear, however, whether that transformation will be pluralist or solidarist in nature.

Although I have not examined ancient Chinese conceptions of the human–nature relationship in great depth, that does not mean that they are insignificant to contemporary Chinese climate practices or international society in general. Indeed, I suggest that international relations theorists pay more attention to traditional Chinese philosophy, the ideas of which could spur innovation among scholars. For instance, Kubin (2010, 524) suggests that, in the global era, the 'Chinese view of *natura naturans* and the Christian concept of *natura naturata* could complement one another so that nature becomes both the real and the spiritual home of man'. Similarly, Daoism and Buddhism could provide fruitful ideas for innovative environmental practices both locally and globally. However, for the time being, they seem to have had little effect on Chinese attitudes and behaviours towards nature and are unlikely to increase their popularity in Chinese political spheres because they form 'too radical an alternative to the Communist Party's statist tradition of remolding nature' (Shapiro 2001, 214). By contrast, Confucianism's anthropocentrism could be 'well tempered through incorporating a Daoist sense of humility and understanding of humans as part of nature, an approach articulated by a surprising number of educated young Chinese' (ibid.).

From the perspective of climate responsibility, an important question is whether China's new notions of responsibility and its philosophical traditions will exert genuine global influence on how great power responsibilities are defined or provide the world with new ideas about humankind's relationship with nature. At present, I remain highly sceptical of China's potential to effect the mentioned fundamental change in values in international society. China has insufficient soft power to generate such change because its autocratic governance system and the recent decision to elevate President Xi Jinping as a lifelong leader of the state raise considerable suspicions and fears worldwide. Such concerns urge the renewal of US leadership in international climate politics. As an established great power, the United States needs to fulfil not only its pluralist great power responsibility to maintain international order and security by taking global climate security challenges seriously but also its solidarist great power responsibility to advance human wellbeing globally by using diplomatic tools to promote the efficient implementation of emissions reductions around the world. Without ambitious great power leadership, it is unlikely that international society will manage to prevent dangerous climate change from happening – and that would be the ultimate tragedy for all humankind.

Notes

1 Cf. Bull (1980).
2 However, as Buzan and Cox (2013) demonstrate, the rise of the United States and China show interesting similarities.

Bibliography

Bull, Hedley. 1980. 'The great irresponsibles: The United States, the Soviet Union, and world order'. *International Journal* 35, 437–447.

Buzan, Barry & Michael Cox. 2013. 'China and the US: Comparable cases of 'peaceful rise'?'. *Chinese Journal of International Politics* 6, 109–132.

Callahan, William A. 2008. 'Chinese visions of world order: Post-hegemonic or a new hegemony?'. *International Studies Review* 10, 749–761.

Chen, Zhimin. 2009. 'International responsibility and China's foreign policy'. In Masafumi Iida: *China's Shift: Global Strategy of the Rising Power.* NIDS Joint Research Series no 3. Tokyo: The National Institute for Defense Studies, 7–28.

Falk, Richard A. 1972. *This Endangered Planet: Prospects and Proposals for Human Survival.* New York: Vintage Books.

Falkner, Robert. 2012. 'Global environmentalism and the greening of international society.' *International Affairs* 88:3, 503–522.

Falkner, Robert & Barry Buzan. 2018. 'The emergence of environmental stewardship as a primary institution of global international society'. *European Journal of International Relations.*

Jackson, Robert H. (1996): 'Can international society be green?'. In Rick Fawn & Jeremy Larkins (eds), *International Society after the Cold War: Anarchy and Order Reconsidered.* Basingstoke: MacMillan Press, 172–192.

Jin, Canrong. 2011. *Big Power's Responsibility: China's Perspective.* Translated by TuXiliang. Beijing: China Renmin University Press.

Harris, Paul. 2010. *World Ethics and Climate Change: From International to Global Justice.* Edinburgh: Edinburgh University Press.

Kopra, Sanna. Forthcoming. 'China and the UN climate regime: Climate responsibility from an English School perspective'. *Journal of International Organizations Studies.*

Kopra, Sanna. 2018. 'China, Great Power Management, and Climate Change: Negotiating Great Power Climate Responsibility in the UN'. In Tonny Brems Knudsen & Cornelia Navari (eds), *International Organization in the Anarchical Society: The Institutional Structure of World Order.* New York: Palgrave Macmillan.

Kubin, Wolfgang. 2010. 'The myriad things: Random thought on nature in China and the West'. In Hans Ulrich Vogel & Günter Dux (eds), *Concepts of Nature: A Chinese-European Cross-Cultural Perspective.* Leiden: Brill, 516–525.

Medeiros, Evan S. 2009. *China's International Behavior.* Santa Monica, CA: RAND Corporation.

Palmujoki, Eero. 2013. 'Fragmentation and diversification of climate change governance in international society'. *International Relations* 27:2, 180–201.

Shapiro, Judith. 2001. *Mao's War Against Nature: Politics and the Environment in Revolutionary China.* New York: Cambridge University Press.

Speth, James Gustave. 2008. *The Bridge at the Edge of the World: Capitalism, the Environment, and Crossing from Crisis to Sustainability.* New Haven, CT: Yale University Press.

Victor, David G. 2011. *Global Warming Gridlock.* New York: Cambridge University Press.

Wang, Min. 2011. 'Statement by H.E. Ambassador Wang Min, Deputy Permanent Representative of the People's Republic of China to the United Nations, at Security Council's open debate on maintenance of international peace and security: Impact of climate change'. Accessed 4 March 2018. www.china-un.org/eng/chinaandun/economicdevelopment/climatechange/t849980.htm.

Williams, John. 2005. 'Pluralism, solidarism and the emergence of world society in English School theory'. *International Relations* 19:1, 19–38.

Xie, Zhenhua. 2010. 'Speech at the high level segment of COP16&CMP6'. Accessed 23 October 2016. http://unfccc.int/files/meetings/cop_16/statements/application/pdf/101208_cop16_hls_china.pdf.

Index